ENGINEERING LIBRARY

D1521292

Reorganizing Data and Voice Networks

Communications Resourcing for Corporate Networks

For a complete listing of the *Artech House Telecommunications Library*,
turn to the back of this book.

Reorganizing Data and Voice Networks

Communications Resourcing for Corporate Networks

Thomas R. Koehler

BOSTON | LONDON
artechhouse.com

Library of Congress Cataloging-in-Publication Data
A catalog record of this book is available from the U.S. Library of Congress.

British Library Cataloguing in Publication Data
A catalogue record for this book is available from the British Library.

Cover design by Yekaterina Ratner

First published in the German language under the title "Netzwerk-Konsoldierung" by Addison-Wesley, an imprint of Pearson Education Deutschland GmbH, München.

International Standard Book Number: 1-59693-040-3

10 9 8 7 6 5 4 3 2 1

Contents

1

The Changing Face of Communication

> It is not the strongest that survive, not the most intelligent, but those that are the most responsive to changes.
>
> —*Charles Darwin*

A great deal has been written, talked, and argued about the role of technology as a driving force for change. The spectrum of publications in this area can hardly be overlooked and ranges from purely technical books to publications focusing on the effects on society to essays on the history of technology. But, apart from purely technical documents, publications dealing with voice and data networks are rare. Therefore, the main theme of this book involves the consequences of technological developments and change on the corporate infrastructure. Companies must face the revolution triggered by technological progress. The impact on corporate infrastructure carries not only risks but also chances to reorient and strengthen the competitiveness of your own organization.

1.1 Communications Technology as an Enabler and Driver of Development

The development of information and technology has already changed the world and continues to do so, possibly even faster than we expect. The "world" in this sense means not only the environment for every individual but also for every corporation. If you analyze this change and abstract from single applications or products, only a few core driving influences are seen to be shaking up our world: the rising bandwidth of existing network connections, the expansion of networks throughout space and time, and the convergence of networks and applications. For corporations, these technological influences allow the company to expand beyond the boundaries of the "bricks and mortar" of headquarters and branch offices and also beyond the typical business hours.

1.1.1 Bandwidth Increase

How much bandwidth is enough? There is no final answer to this question, which is similar to the question about how much computing power is enough. The amount of data that has to be moved through local-area and wide-area networks is still increasing year after year even though general statements made during the era of Internet hype—such as "data traffic will double every 3 months" or "Gilder's law (1997): Available bandwidth used for data transfer will need to triple every year"—turned out not to be true in the long run.

Growth is still strong growth, however, in the area of, for example, wide-area Internet Protocol (IP) traffic [1]:

- International IP traffic rose by more than 67% in 2003 alone.
- A heavy increase in peak transatlantic data traffic was seen: from about 100 Gbps in 2003 to an expected 466 Gbps in 2006.

A similar development has also taken place in household Internet connections. The trend toward higher bandwidth and higher usage is unbroken. Internet access via broadband (cable and DSL) is still on the rise, while total market penetration of Internet access is approaching the saturation point. This means that dial-up connections are losing ground against broadband connections almost everywhere in the developed world. Company connections to the Internet and connections within their own wide-area networks (WANs) are also still growing strongly.

We are also seeing a costimulation effect between the increase in computing power and the increase in bandwidth. For local-area networks (LANs) one can see a pendulum-like trend leading to growth in quantum jumps (from 10 to 100 Mbps and now to 1,000 Mbps). Within corporate networks, bandwidth has increased within the LAN as well as in the WANs but at different growth rates. There are even signs that growth in network connection bandwidth has exceeded the growth of computing power [2].

A major driving force for network growth is the single personal computer (PC) workplace and the applications used with such a setup. The necessary bandwidth per PC is growing as a greater number of increasingly complex applications that require high-bandwidth connections to run smoothly are used on the network. About a 30% to 40% per year increase in bandwidth demand is being driven by applications' needs says Salomon Smith Barney in an internal study.

Technological advancements such as automatic software updating and patching, centralized data storage, and even remote backup systems also lead to higher bandwidth demands.

It is not only demand that drives this growth, but also the availability of bandwidth that leads to higher consumption. A few years ago, most Internet-savvy users did not even think about sending large e-mail attachments without first asking if it was appropriate. Now it is quite normal to send megabytes of PowerPoint and Portable Document Format (PDF) files to whoever might be potentially interested in them. Even e-mail newsletters nowadays often come with several megabytes of piggybacked data. This behavioral change cannot be explained simply by looking at the large proportion of Internet newbies; it also is driven by those who have been using the Internet for a decade or longer.

Similar developments can be seen in the increasing size of Web-based applications. In the early days of the Web, most developers and designers of Web pages focused on reducing the amount of data that had to be transferred to display a Web page by shrinking the pictures and leaving out unnecessary parts, which reduced the loading time. There even was a huge discussion, culminating a few years ago, about the maximum amount of time a user would be willing to wait in order to view a particular Web page. It now seems as if nobody cares about this anymore. Web site content is getting richer and bigger, leading to unacceptable loading times on dial-up connections and thus stimulating further broadband growth.

Even portal Web sites used for employee services suffer under visual and traffic-related opulence and this also helps to drive the demand for bandwidth. Generally speaking, most of the transmitted data traffic within and outside of corporate networks is not what we would call content (such as financial data or the results of a database query), but rather ancillary information designed to make the content look good and work well on the target system.

Several other technological advancements also stimulate bandwidth growth. For example, thin clients are typically devices without hard drives that rely heavily on centralized servers for their computing and that obtain applications and data over the network.

But there is another side of the same story, which is sometimes called the bandwidth paradox. Compared to classical telephone networks, data networks are typically less busy (much more below capacity). The University of Minnesota came to the conclusion that private data connections (such as the ones used between company branch offices and within corporate networks) are typically only driven at 3% to 5% of capacity, and Internet backbones at 10% to 15%, whereas the voice telephone network averages 30% of capacity used [3]. What seems weird and illogical in the first place can be understood and illustrated with detailed knowledge of typical data network usage patterns, as discussed next.

1.1.2 Universal Access

Bandwidth means nothing without availability, but availability describes more than just an aspect of a service agreement, defined as the percentage of time during which a system is up and running and therefore available. Availability also refers to the spatiotemporal possibility of access (to a network). The range of coverage includes a spatial component (range) as well as a temporal component ("always on"), the combination of which has brought massive changes in the way corporations do business. This has evolved with stealth and has often only been recognized by the unexpected effects it brought with it.

Range

One major component of what can be called universal access is range, the spatial availability of the network or, technically speaking, network coverage. Range is one of the major important factors discussed in this book.

Range has developed astoundingly fast during the last few years. The ubiquitous availability of network access—supported by remote log-in, mobile data services, and software that builds and secures the connection to the infrastructure of the corporate headquarters or branch office from anywhere—has helped shift and break the boundaries of the corporation. From a real estate–oriented view of the corporation that was typical just 20 years ago, we are now evolving toward a new understanding of what a corporation is and does. The corporation now reaches into the private homes of employees working part or full time from home, and to mobile workers in hotel rooms and airport lounges. Apart from remote access that is relatively simple to implement for home workers, the greater challenge lies in providing mobile access. Within the last few years, mobile data has come of age and now offers different connection solutions. (These solutions vary, depending on which part of the world you live in, but UMTS and WLAN are expected to provide common ground for the future.) The range of the network is growing and is now almost everywhere within the state, the country, the continent—even the world—wherever mobile phone connectivity is available. With the help of satellite-based data services, this connection is available almost everywhere, even in the so-called "quiet zones" (where you were previously cut off from any connections), such as in the passenger seats of intercontinental flights or in expedition camps in the Himalayas or Antarctica. It was repeatedly reported that there was even a small Internet café/wireless LAN hotspot at the Mount Everest base camp during the 2004 season; it was located in a tent but connected to the world via satellite. One can complain about the loss of this quiet space but this will not stop the process of development. To be out of range, you will now have to go undersea or leave the planet.

Always On

Not only the spatial but also the temporal dimension is needed to make maximum use of a network. Another effect of modern networking is the "always on" paradigm, which replaces the time period where connections to data networks were typically initiated when actually needed.

"Always on" is simple. The network connection does not ever go down and it not ever switched off; it stays permanently on, or at least simulates a permanent connection. The consequence is that one loses latency between action ("I need information on the topic of...") and reaction (result of the query delivered). Besides the impact of bandwidth, which defines how fast something can happen, latency defines the time between the initiation of the process and its actual start, which, from a network point of view, is the beginning of transmission. This is the greatest achievement compared to the dial-up–dominated past: the nonexistent waiting period until a connection is established. Compared to dial-up modem connections, "always on" is another world that leads to different usage patterns, not only in the connections of large organizations but also in connections to homes and home offices as well as in innovative mobile services.

"Always on" also means that messages are forwarded to their destination without hold time. This is especially important for remote usage scenarios. Home office workplaces with "always on" connections allow instantaneous reaction to incoming messages and permanent working with centrally provided applications. This is also true for mobile devices and applications. The most prominent "always on" application for mobiles is e-mail push. With e-mail push there is no waiting for an e-mail client that connects to the mail server and downloads messages after a user-initiated action (the "pull" principle). As soon as new e-mail reaches the server, it will be "pushed" forward to the user device and displayed immediately, closely resembling the short message service (SMS) used in Global System for Mobile Communications (GSM) mobile phones. Sometimes this e-mail push is just a simulation as is true, for example, when an automatic request for new e-mail is originated by the client repeatedly every few seconds; sometimes it is more intelligently controlled using parameters such as time of day, weekday, and the number of messages typically arriving within a certain time frame. In the end this approach comes close to real-time push.

1.1.3 Convergence

There is much talk about convergence in the context of networking discussions. From a marketing point of view, the word *convergence* is already burned out, but for us this is just beginning to become important.

Convergence (from the Latin word *convergere*, meaning "to tend toward something") means "approximation" but also "matching" (of goals or possibilities). A lot of specialist areas outside the world of information technology and telecommunications have their own special view of the term *convergence.* Mathematics, for example, sees convergence as sign of the existence of a boundary value, the convergence of sequences. In ethnology the occurrence of similar cultural habits among folks who are living without any contact with each other is also named *convergence.* There are many other uses of the word *convergence,* such as in the technology of cathode-ray tubes where convergence is used to describe the correct focusing of the base colors red, green, and blue on the screen of a computer monitor or TV set.

For this book, and for us, *convergence* has a different meaning. The convergence of services, most importantly voice and data on a common network, is one of the core aspects and is an important basis for further discussion. Besides the convergence on the network level, convergence on the application level is also of importance. Voice applications converge with screen-oriented business applications and also application front ends move to the Web browser as a universal access tool.

Just a few years ago, much of the convergence-related discussion could have been subsumed in just one word or acronym: *TIME markets* (sometimes also called *TIMES*):

- Telecommunications;
- Information technology;
- Multimedia (sometimes the "M" also stood for mobility or mobile data or mobile services);
- Entertainment (sometimes the "E" was used for e-commerce).

The "S" in TIMES was then used to add the topic of "security." The term *TIME(S)* has almost disappeared in public discussion, partly because it was overused and nobody wanted to hear or read about it anymore, but more importantly because a lot of this convergence between markets has already happened or is still happening. Similar acronyms describing similar effects appear from time to time in special interest publications.

"Triple play," for example, is not only a move in a baseball game but also the term used to indicate the combination of data, voice, and video/TV services, and it is being promoted as a business model of the future. The triple-play concept is aimed primarily at the end user.

Voice and Data

For the future development of information technology and telecommunications within the corporation, the convergence of voice and data on a shared network platform is the central and most important precondition. This convergence development has massive consequences for corporate networks of a single branch office or headquarters (that is, the LANs) or among the different locations of the corporation (that is, WANs). Voice becomes just another application within the data network. Voice-data convergence as the basis for further discussion is described in detail in Chapter 4.

Web as the Universal Application

Convergence in the information technology (IT) and telcommunications (TC) sector is not limited to voice-data convergence. Although this of course plays the most prominent role, application convergence is also important. Within this, the Web plays an important role because it brings together and integrates different applications under a consolidated front end, accessible via a Web browser. Company and employee portal sites show the way and typically have standard Web interfaces. Besides this, e-mail messaging and service portals provided by IT service companies, Internet service providers (ISPs), and telephone companies (telcos) are migrating in the same direction.

Application consolidation under the common roof of a standard Web interface strongly helps to explode the boundaries of the corporate network: There is no longer a special device [e.g., PC, notebook, personal digital assistant (PDA), smart phones] with a special application for every task. Access to the corporate infrastructure, data, and applications is now available to any authorized person with a Web-enabled device. This device could be, for instance, a PC at a customer's site, a computer terminal in an Internet cafe, a public service site, or a portable device linked to a hotspot somewhere. The basic precondition is only that the user has access to a Web browser using the common standards. This can also be a Web browser in a mobile phone or in the navigation system of a high-tech automobile. The Web as the "universal application "strongly increases the availability of corporate software applications.

Of course, this does not mean that this type of access is desired or even supported by individual corporations, because any outside connection will indeed have inherent security risks and therefore raise support and service issues. The types of possible security problems, such as stealing passwords, listening to sensitive data transmitted over the network, and so-called "man in the middle attacks" in which an attacker tries to manipulate the connection for his or her own needs, are extremely varied. Security is a major issue but we will not discuss it here. The power of universal access is so strong that it cannot be stopped. It will proceed despite the security-related concerns. Employees will simply search and find new ways to make use of this potential. A network administrator at a medium-sized company who is summoned during his holidays to take care of a server problem that arose while he was not at his office desk will probably intensely dislike having to drive to his office to fix the problem if there are ways of doing this remotely, maybe via mobile phone or via an Internet terminal found in the nearest Internet café. If he is an employee with special knowledge, it will probably be more important to keep him happy and secure his abilities for the corporation than to let potential security concerns rule out the possibility of remote access for him.

1.2 Consequences of These Changes on the Individual

If one at first ignores the aspects relating to the corporation itself and focuses solely on the individual, one major observation can be made: The once strict distinction between work and leisure time seems nowadays almost totally abolished.

Until a few years ago, only top management and self-employed individuals were known to be always working, including working overtime whenever it was needed, largely at cost of their leisure time. Besides this, one occasionally met an employee in standby mode, during which he had to move to his desk from wherever he was and whatever he was doing, within a defined short period of time, to deal with some type of emergency notification. For all other employees, working life ended with leaving the corporate grounds.

Nowadays large parts of the workforce experience a mixture of work and leisure time: Work is done partially at home after working regular hours at the office or, instead of working at the office, e-mails are answered from home, while remaining mobile any time of the day or night. The office phone is forwarded to the mobile phone or home phone. For callers, it is not a rare event that, after calling an office number, the answer comes from someone who actually manages his or her leisure time or takes care of his or her children but nevertheless is all-ears for business affairs.

This new role model, which is often an unwritten contract between employer and employee, is taking possession of leisure time. Note, however, that managing one's own private life now sometimes occurs during business hours: Making a phone call to a tradesman who is working at a private home, or checking in on an Internet auction that is due to end on Tuesday afternoon, demands attention during working hours. Leisure time is finding its place within the business world of most corporations. No assessment as to whether this is a good or bad thing should be made at this time. The underlying principle of cause and effect is important for us in this book because it is relevant for the planning the future of corporate network infrastructures.

The availability of supporting technology has made these types of changes possible. The wide dissemination of mobile phones, remote access gateways, and broadband home connections are the key driving influences of this development. Business communications is moving from the domination of telephony and faxes toward heavier e-mail usage that is largely spatiotemporally decoupled: "spatial" because e-mail does not require a special place to be received, unlike traditional fax services, which are bound to a fax machine somewhere, and "temporal" because e-mail does not require immediate attention and an immediate response. Therefore, e-mail helps in combining leisure time with gainful employment.

This maceration of the boundaries between work life and leisure time can also be illustrated by looking at some of the side effects. In the last few years, notebooks have replaced more and more desktop PCs. This is not only true for private households but also in corporate environments, even when workers—unlike management, salespeople, and service people—have no real need for mobility and stay at the office all day, all year while working for a particular employer. The falling prices of notebook computers support this development. The disadvantages of notebook computers compared with desktop PCs, which include lower computing and graphic power and less storage capacity (at least if one compares PC systems in a similar price range,) still exist but are unimportant for typical office work. This also helps to make the transition.

A 2004 study from the analysis company Meta Group shows that this transition is already sufficient to boost usage times [4]. Weekly usage within the United States will increase by about 6 hours per week. Meta Group also reports that the same trend in hardware supplies in Europe will lead to 4 hours and in Asia to 12 hours of additional usage time.

Even if one does not believe that all of these additional working hours will transform directly into business gains, one must admit that probably at least part of the additional time will be used for business purposes. Usage while traveling to and from the workplace, but also usage at home, during what was previously leisure time, will at least in part be done to fulfill business needs and will make employers happy. Private usage during leisure hours can be easily accepted if you calculate the additional hours you will not have to pay as an employer. But one main problem remains: Security problems can arise through private usage, for example, with the installation of unwanted and potentially malicious software. Spyware can be a serious problem, even if the employee is acting in good faith, and IT department must be aware of the security risks inherent in the use of corporate notebook computers.

Additional positive effects can be expected from smart phones and handheld PCs. Handhelds in particular, which allow access to corporate data and which have integrated messaging capabilities, are generating more and more interest. A pioneer in this area of development was the Blackberry mobile e-mail device of RIM, a Canadian company. RIM not only supplies hardware and software for using e-mail everywhere but also the necessary arguments to underline its business value. A study done by the consulting company Ipsos Reid on behalf of RIM shows significant direct cost savings and productivity increases but also more of what they call "immediacy" [5]. Their definition of immediacy gives a financial value to timeliness and calculates the cost savings achievable through shorter response times in business-critical correspondence. They argue that fast decisions are worth more than slower decisions and calculate a type of return on investment (ROI). Of course, these are only assumptions on what it is concretely worth for an e-mail to be answered 30 minutes earlier.

Whether one actually believes or does not believe the actual numbers presented here or in similar reports is irrelevant to our discussion. What is more important is that notebooks and handheld devices help to make use of the times when we were previously passive, since there was nothing else to do, such as sitting in a train or waiting at an airport for business purposes. Notebooks and handheld devices therefore are important tools for supporting this change.

For the larger part of the working population, the understanding of work is changing from the usual "9 to 5" model to a new model, in which someone feels responsible for a certain task or project not 24/7 but whenever it is necessary.

From the corporate viewpoint, the direct consequences of the disappearance of the spatial and temporal borders of the enterprise are increased productivity and faster response times but also potential security hazards. Finally, this new way of working imposes additional requirements on system support, including the provision of remote access services as well as support and security for those who are outside the spatiotemporal borders but are working with and for the corporation.

We cannot close Pandora's box and it is impossible to turn back the clock. The pressures of competition force us to become increasingly flexible in the way we do business.

References

[1] Primetrica Global Internet Geography 2004, http://www.telegeography.com/products/gig/index.php.

[2] *Informationweek*, No. 25, April 4, 2003.

[3] "Data Networks Are Lightly Utilized, and Will Stay That Way," *The Review of Network Economics,* Vol. 2, No. 3, September 2003.

[4] Meta Group, 2004.

[5] Ipsos Reid, "Blackberry Return on Investment," 2004.

2

Status Quo of Corporate Voice and Data Networks

Facts do not cease to exist because they are ignored.
—*Aldous Huxley, (1894–1963)*

Where do YOU stand today? Where do WE stand today?

It is useful to take a close look at how corporate voice and data networks currently function and the technologies on which they rely. As we do this, we should also examine the local infrastructure, not only of the corporate headquarters and branch offices, but also the corporate WAN, which includes the infrastructure among the branch offices. This should be done before planning any network strategy changes.

2.1 Corporate Voice and Data Networks

Attempting to establish a typical real-life scenario within the pages of a book can be risky, because there is the danger that the reader will not be able to identify himself or his company with the situation described. But companies that watch the market developing and spend time analyzing it, as well as telcos and suppliers of network infrastructure, generally agree with the usefulness of this approach.

The network infrastructure of a corporate headquarters or branch office usually consists of two distinct networks: one for data and one for voice services:

- *Voice network.* A typical corporate voice network can be described as follows: One centralized telephone switchboard connects desktop phones (and cordless handsets) to each other and to the public telephone network (outside lines). The number of outside lines has a close relation to the number of phones within the location. This number differs from industry to industry, for example, call centers require many more outside lines than an average manufacturing corporation having the same number of desktop phones. This number is usually derived from industry experience or from previous experience. A number of different calculation schemes exist that help with deciding on how many outside lines are needed. Dedicated cabling, or at least a nonshared use of parts of the cabling in newer installations, is also typical.

- *Data network.* The state of the art in corporate data networks is an IP-based network among the workstations, or between servers (which store data and applications and offer these to the network clients) and workstations (often called *clients*). This network also integrates the capability of transmitting legacy data between applications and systems. As long as this network is located on only one corporate site, it is referred to as a LAN.

Distinct cabling for the data network is typical. Besides the cabling, which is standardized into different categories (Cat-3, Cat-5, and so forth), additional equipment is needed for connecting network segments to each other or to the outside world. Components used include routers, hubs, and switches. Server systems are sometimes also included into the enumeration of network equipment.

In addition to the wired network, the last few years of technical development in networking have brought us the wireless infrastructure. A wireless local-area network (WLAN) is an extension of a fixed network (see Figure 2.1). Few companies rely solely on WLANs for their networking, so most installations are single cases used for special areas such as company warehouses, conference rooms, hotel lobbies, and so forth, and are not roaming enabled.

All equipment based at this company location, whether it is wireless or hardwired to the LAN is normally also connected via wide-area connections. Normally only one point in the LAN has a connection to the outside world, and it allows all clients who have permission to access to company branch offices and public networks such as the Internet and the public switched telephone network (PSTN). This scenario is illustrated in Figure 2.2.

2.1.1 PABX Within the Corporation

The core of most corporate voice implementations is the PBX or PABX (private branch exchange/private automatic branch exchange), also known as the

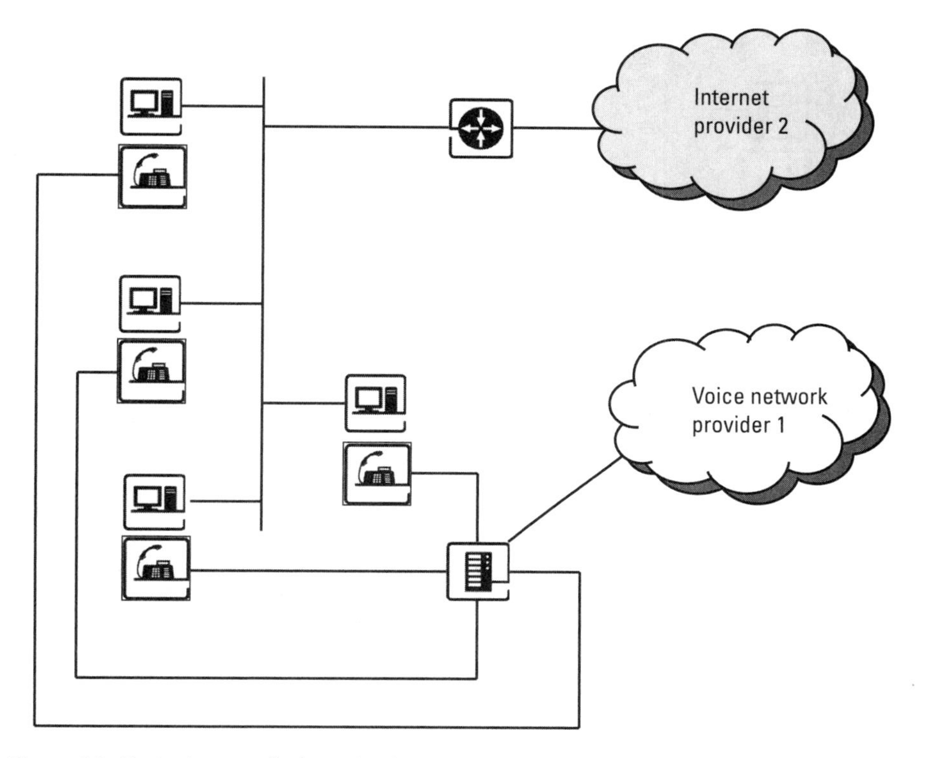

Figure 2.1 Typical constellation of voice and data networks in an enterprise (simplified representation).

company switchboard. This connects the devices (desktop and cordless phones, fax machines, data modems, and so forth) within the company site to each other and also to the PSTN).

In certain countries, including the United States, market development also led to a service-oriented system being offered, the so-called CENTREX (central exchange) service. With CENTREX, the telephone network provider offers telephone exchange services to companies by using equipment that is located within the provider network and not on the customer's site. This equipment is also serviced and supported by provider organization employees. In this scenario, there is no or little equipment apart from the phones and fax machines at the customer's site. CENTREX services are popular in the United States and in several other countries, but the dominating model of corporate phone service delivery in the Western world is still the PABX, located at the company site or branch office.

For decades the development of PABX systems did not keep pace with IT development. PABX has had much longer development cycles than IT

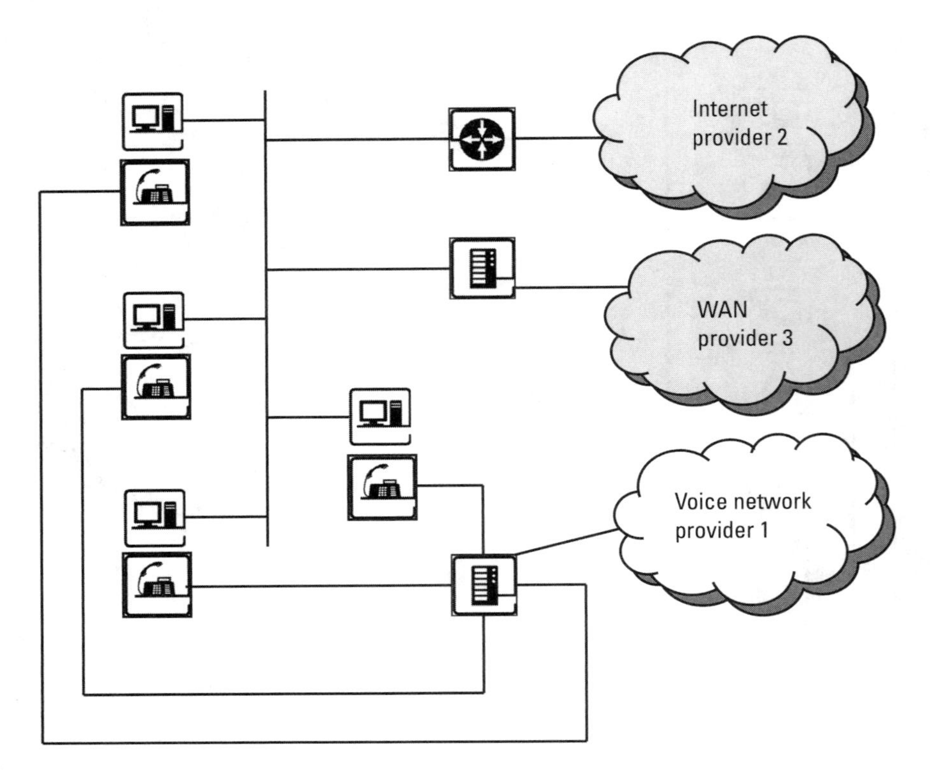

Figure 2.2 Typical constellation of voice and data networks in an enterprise with WAN connectivity (simplified representation).

components. The feature set has steadily increased over the years, but not at the same speed as in the IT industry.

A few major players have dominated the PABX market in the United States and Europe. Each of them sold, and still sells, its own proprietary equipment to companies of all sizes. Their typical business model can be described as leasing or rental of a certain PABX system for a time period of several years based on a setup/configuration defined and implemented at the beginning of the lease. The contract typically ran for a 3- to 5-year period and, until the 1990s, even 10-year leases were not uncommon.

The user was stuck with this starting configuration over the time frame defined in the contract. If the configuration needed to be changed, the user had to pay high additional fees for those changes or when moving offices. Lower usage or a need for fewer telephone outlets, for example, due to the company shrinking or the closing of a branch office, did not usually lead to lower monthly payments because, once agreed to in the contract, the setup usually had to be paid for over the complete lifetime of the agreement.

Getting out of the legal claws of such a contract was typically not possible or economically viable. Third-party changes and additions were also not an alternative because these were usually excluded in the terms of the contract or were simply not possible due to technical restrictions, proprietary protocols, or missing access to, and understanding of, the software interfaces or the configuration software itself. In most cases, acceptance of the terms and financial conditions of the original supplier was and is the only way to implement changes.

To sum this up, one can say that for several decades typical "lock-in" effects used industry-wide by PABX manufacturers and network operators acting largely as resellers of PABX systems.

Things look different if one takes a closer look at the underlying cabling infrastructure. A typical scenario for this is as follows:

- Local service companies usually do the cabling. They rely on the planning of the PABX companies and implement the necessary cabling on this basis within the company facilities (owned or rented).
- Inside office buildings built during the last few years, basic cabling for TC systems is widely included and is seen as part of the rental object.

Most connections from the internal telephone network to the PSTN rely on fixed lines. However, options are available for building a connection over the "last mile" via wireless links or even via satellite connections. This depends largely on regionally specific service provider offerings.

Facility management of office buildings sometimes includes not only prewired rooms but also a full telephone service, including the switchboard as well as all necessary equipment such as the desktop phone itself. These service offerings can often be found in buildings and office centers targeting young or start-up entrepreneurs or companies who are only there for limited periods of time such as several months, which makes any planning and implementing of own equipment economically unreasonable. The quick start option is another major advantage of this concept.

A similar approach is the CENTREX service mentioned earlier in this chapter. Compared with traditional phone services, CENTREX services have several advantages:

- More flexibility;
- Simple reconfiguration;
- No limits in adding connections/phones;
- Easy integration of additional branch offices and teleworkers (no limitation of the dialing plan to one location only);

- Lower downtime (risk);
- Lower capital expenditure;
- Cost savings (depending on the tariff model);
- No space needed for telecommunications equipment.

Apart from the service-oriented offerings already mentioned, the typical scenario described here applies to the corporate reality of most parts of the Western world.

The relatively slow development of company-centered PABX systems has had special advantages: The product quality is relatively high. Downtimes for PABX systems are rare in comparison with everything IT related. The telephone users within corporations are accustomed to the same high quality as any user of the public telephone system, with no or only little service downtime.

Looking at the costs of telephone services, one can say that the costs are relatively transparent to most corporations, because large parts of the system are externally provided. Although consolidation of services has been planned by most corporations for several years now, purchasing of telecommunications services is still very often done by an organizational unit that is separated from the IT department, or is at least based on different budgets. It is not easy to determine how big the percentage of the corporations is that still act independently from IT when buying phone services, because analyst opinions vary widely on this topic, as the author of this book discovered during research in 2005.

2.1.2 LAN Within the Corporation

Cabled LANs are cabled in much the same manner as telephone systems but aside from this there are few similarities. The most important differences lie in the equipment used (routers, switches, servers, patch panels, and so forth). Companies of all sizes typically buy their network equipment rather than renting or leasing, and management is usually done in-house.

Because a single piece of network equipment usually only costs a relatively small amount of money, leasing is not viable due to the overhead costs. A larger proportion of leasing can be found in storage and server systems but the corporation also still owns most of these.

Software licenses for the applications necessary for providing network-based services are also usually purchased and not leased or rented. Suppliers providing a leasing model or a "pay per use" pricing scheme have experienced massive problems with corporate customer acceptance.

The LAN world revolves around companies providing services by themselves using their own equipment. Exceptions to this rule of thumb can only be found in external data connections. The device necessary for connecting to the outside world is very often owned and serviced by the connectivity provider. Providers have a special name for this: customer premises equipment (CPE).

Quality varies between LANs because most of the installations are the products of ongoing development over several years and are therefore based on components from different manufacturers and technical generations. Service and quality agreements within a self-managed LAN are problematic. As a consequence, there are normally no agreed-on quality and service standards. Within self-managed corporate LANs, quality is usually only checked and provided as a statistic—after the fact.

Besides the heterogeneous infrastructure, other factors are working negatively on the quality of service that can be achieved. Most small and medium-sized companies are not organized in a three-shift system, so there are few or no service personnel available outside the normal 9-to-5 working hours. If network or computing problems arise, support becomes a question of the time of day.

Small and medium-sized corporations typically fulfill these needs by using automatic service messaging, for example, when an incident occurs an SMS or e-mail message is sent automatically by a surveillance system to the administrator at home. If any important part of the infrastructure goes down (e.g., switch, server, router), this incident is automatically detected and the information is forwarded to the administrator, who then solves the problem via remote access or by simply driving to her workplace and taking the required corrective measures.

With regard to network personnel, most corporations suffer from the so-called "guru syndrome." Only a few people have detailed knowledge of the complexity of the company infrastructure and know what to do when an incident arises. This problem is often underlined by missing or outdated incident plans, leading to even more dependence on the wisdom of the knowledgeable few.

Typically, the IT department is also responsible for networking, or at least for data networking. A separate department may be responsible for the network but this is not always the case. The IT department also typically administers the budget for the network infrastructure. The total cost of this "trial-and-error networking," however, is often not calculated or is only calculated for certain parts. Smaller companies in particular pay little regard to their networking infrastructure. Investments in new equipment are made within these groups typically only as a consequence of severe problems or as a response to downtime incidents.

2.1.3 WAN Within the Corporation

WANs contain the infrastructure necessary to connect branch offices and other company sites with each other and with the company headquarters, on a national and international scale. External suppliers, for example, large carriers, typically provide these connections as rental lines.

Most WANs are initially deployed using a "hub-and-spoke" architecture, in the shape of a star. Several lines lead to other company sites from the corporate headquarters in the middle. If most service deployment is concentrated in the corporate headquarters, with most communication occurring between branch offices and headquarters and only a small amount between the branch offices themselves, then this network structure is the architecture of choice.

Corporate WANs are based on frame relay, asynchronous transfer mode (ATM), rental lines, or a combination of these. Cost-oriented companies also often use the public Internet to connect to their branch offices via an encrypted connection. The Internet connection is already there, so why not use it to connect to any other branch or outlet? Apart from the security risks, the problem with this approach lies foremost in the lack of quality of service. The use of the public Internet provides no guarantees. If network traffic is high, then delays must be expected. These connections might be good enough for the implementation of connections to home workers and smaller branch offices, or to provide extranets to suppliers or customers, but they are questionable when it comes to important applications.

Corporate WANs also include Web activities such as corporate Web sites or e-commerce systems. A corporate Web site is now standard for companies of all sizes. Almost every big corporation has had a Web site for several years, while most small and medium-sized businesses also offer information on the Web, except for a few local or traditional ones that still refuse to have a Web site or stopped having one after they discovered that the value they received from the site did not justify the costs.

E-commerce and e-business systems for Internet-based transactions, extranets for supplying connections to suppliers and customers, and development and production partners add to the picture.

The connection to the PSTN is provided by one or several carrier companies and usually exists individually at every company location. For telephone service between the locations, special offers provided by the carriers are often used that provide attractive conditions for these types of connections as well as added functionality such as short dialing. In newer installations, telephone service is often also provided over the data lines of the WAN. Although this can already be seen as a sign of the convergence of voice and data, the two systems are still very often based on different budgets. If an additional line connection becomes necessary, for example, for a new branch office or as a consequence of a merger between corporations, most decision makers tend to select the provider offering the best price for the extra connection. In the long run, this leads to a WAN consisting of lines provided by several suppliers with different contract durations, different quality-of-service (QoS) agreements, different support phone numbers, and so forth. This, in turn, leads to management problems, and problems if an incident happens, because it is often not clear which of the several line providers must be contacted and—if the selection problem is finally solved—what kind of service agreement is applicable to the particular line. Large carriers often get calls from corporate customers complaining about the loss of service or downtime that are actually addressed to the wrong provider. Sometimes these calls even include threats of service agreement cancellation (which of course never existed with this particular provider for this particular connection, since it is actually serviced by someone else).

Multiple contracts for outside connections not only lead to problems in contract management and when trying to solve technical problems, but also to more administrative work. The administration and auditing of the multiple service bills can be costly. The author knows a number of enterprises where several people work only on the administration of provider invoices. In these cases, if a proper total cost of ownership calculation is done, the savings achieved by buying lines one by one will probably be outweighed by the administrative costs. As a whole, this "single-point optimization" process actually prohibits optimization of the whole.

Infrastructures developed in the manner just presented are not usually consolidated to a single source or a small number of vendors, due to differing contract structures, durations, and payment schemes. In a few cases, changes in the organization (e.g., people leaving for new assignments) lead to the loss of information about provider relationships. As a consequence, difficulties often arise when attempting to consolidate, and the first task becomes finding out which contracts exist with which providers. This problem gets even worse when buying decisions are distributed throughout the organization and the branch offices purchase locally. Under these conditions, consolidation can be a nightmare to accomplish.

Finally, one must state that the hardware necessary for WANs and LANs will typically be purchased, with the exception of devices provided by the carrier as a part of the service.

2.2 Comparing Industries

Not all industries and corporations act in the same manner with regard to network communication. One can see large differences when looking at the voice and data networks of companies from different industries. In this context one also discovers that different industries have different priorities when it comes to voice and data networks.

2.2.1 Manufacturing

Voice and data networks are very important to the manufacturing industry. The increasing need to transfer data within the supply chain, and also within the corporation, makes reliability a major issue. A single long downtime incident can lead to the "death" of the company, especially within the automotive industry, where not only "just-in-time" delivery matters but also "just-in-order" delivery is important, due to the high customization of the final product. A parts supplier must ensure that it gets all of the correct parts to the assembly plant, in the correct order and at the right time, which won't happen without a reliable network infrastructure.

2.2.2 Logistics and Transport

Due to the integration of logistics companies within the supply chain of the manufacturing industry, things are similar here. The requirements are the same.

Logistics companies offering warehouse services and contract logistic companies are being forced to invest in infrastructure in order to satisfy the requirements of their customers.

2.2.3 Banking and Financial Services

Banks and financial service companies have special requirements for their information technology and related infrastructure, due to the regulatory influence of the government. Of course, the regulatory impact depends on the country in which they do business, but all countries of the Western world have some type of financial sector legislation. This is one reason for the redundant, heavily secured systems that are standard within this industry. New technology is adopted slowly. After the e-business hype at the end of the 1990s, with false allocation of millions of dollars, most banks and insurance companies have still not fully recovered and work very conservatively when it comes to investing in IT and telecommunications infrastructure. Naturally, security is also a major issue.

The current discussion in most banks and insurance companies focuses primarily on the correct balance between "do it yourself" and outsourcing. In a study published in September 2003, the consulting company Accenture complains that most players within this industry focus too little on their core business and do too much work on their own with their own personnel that would be better done by service providers [1]. The ability to compete suffers under this scenario. Discussion within this sector, therefore, is often outsourcing related. The "bigger is better" mantra is also often repeated, due to the expected cost savings through higher transaction volumes and is a topic within mergers and acquisitions in this industry.

An interesting question arises in this context. How do larger outsourcing transactions influence mergers and acquisitions? Isn't the main reason speaking in favor of a merger the cost by size issue? Will the aspired-to savings occur if one of the partners already relies heavily on outsourcing and is bound to a long-term contract with a service provider?

2.2.4 Retail Industry and Customer-Oriented Services

Large parts of the retail industry are not up to date when it comes to network infrastructure. Wide-area networking is very often done in the most affordable way: by using the public Internet. The connections to the sales outlets are very often only considered from the cost point of view.

Most point-of-sale terminals can work offline, without any connection to the corporate headquarters except for a daily update. The same is true of rental car company service stations and other service providers. Apart from this, there are, of course, mission-critical applications inside these industries that rely heavily on the network, but these are rare.

2.2.5 Industries and Downtime Costs

A major indicator of the importance of the network infrastructure is the costs that occur as a consequence of a breakdown leading to downtime. Table 2.1 shows examples of downtime costs within different industries in the United States and gives an indication of where the most money will probably be spent on the reliability of the infrastructure to prohibit these breakdowns from occurring. Note that the differences among companies of the same industry are large, so the examples presented here for different companies do not necessarily reflect the whole industry.

Table 2.1
Downtime Cost by Industry and Activity

Activity	Industry	Downtime Cost Per Hour (U.S. dollars)
Brokerage	Finance	$6.45 million
Credit card authorization	Finance	$2.6 million
Catalog ordering	Retail	$90,000
Airline reservations	Transport and logistics	$90,000

Source: [2].

2.3 Company Sizes in Comparison

Differences between industries are not the only important factor. Differences in company size are also important because small and medium-sized enterprises have different approaches to network strategies compared to large enterprises.

Most large companies and multinational companies typically place much effort into the strategic planning of the infrastructure. Decision making is normally centralized with a focus on the whole corporation. Risk minimization is a core factor. ROI/TCO calculations are understood and very often used within the decision-making process. Relying on external service providers is also quite normal and has been done for several years.

The approach to consolidation is typically appropriate. Small and medium-sized companies very often engage in more short-term thinking than larger ones and have fewer strategic considerations when it comes to networking. Departments very often have their own decision makers and less centralization is therefore the norm. Cost savings are a key factor in decision making. ROI and TCO calculations are widely understood but seldom accurately used. There is very often little background knowledge available on this topic. External providers are only trusted in certain areas. The adoption of new technology is low and a trial-and-error approach is often used.

2.4 Branch Offices—A Special Case

From a historic point of view, branch offices were often treated as stepchildren when it came to IT and telecommunications. One consequence of this was that these branches very often began buying technology and services locally. This often led to compatibility problems and inconsistencies within the corporation as whole and, consequently, to higher costs and less effective communications. From a holistic viewpoint, another consequence can be seen in system redundancies, since not only the headquarters but also one or several branch offices have independent solutions for one and the same problem.

Typical branch office problems are as follows [3]:

- Delays due to missing/insufficient access to central applications;
- Delays in decision making and problem solving due to inconsistencies within the communication systems between headquarters and branch offices;
- Higher support and service costs;
- Higher TCO for the local infrastructure and less functionality;

- Problems in using branch office personnel resources for headquarters tasks (e.g., for implementation of temporary additional call centers);
- Information losses due to technical incompatibility;
- Absence of user training and technical support for systems that are only used at one or a few branch offices but not in the headquarters;
- Feelings of isolation from the rest of the corporation, from headquarters, or from the customers.

As more interaction is required between headquarters and branch offices and the software systems become more centralized, these inconsistencies can no longer be tolerated since they not only compromise the productivity of the individual worker but the productivity of the entire organization as well. Efforts to break out of the branch office dilemma focus primarily on consolidation and standardization of the corporate infrastructure while centralizing services at the same time.

2.5 Requirements for Consolidation

In summary, one can state that most companies use more than one networking platform in parallel and operate these by themselves. Operation, service, and management of these platforms tie down valuable resources.

Applications are very often provided at several locations, and services are very often not consistently defined throughout the entire organization. Apart from these general statements, it is not easy to find hard facts supporting the need for consolidation, since most corporations do not want to release actual figures and instead treat them as trade secrets. The best one can usually obtain are industry-specific key figures for IT spending and infrastructure basics delivered by industry analysts or consulting companies and these are mostly based on assumptions. DHL/Deutsche Post World Net is a major exception. By making PowerPoint slides from the presentation of the chief information officer publicly available, they show how great the consolidation needs are at DHL/Deutsche Post. The following infrastructure numbers were presented:

- Three corporate networks;
- Three global data centers;
- 18 regional data centers;
- 140 firewall systems;
- 290 server locations;
- 1,500 applications;

- 2,500 databases;
- 2,600 business application servers;
- 4,000 IT specialists and IT managers;
- 60,000 e-mail accounts;
- 130,000 desktop computers.

The annual budget to provide this infrastructure is $1 billion. Voice components are not integrated within these budget calculations and are separately calculated. By checking this list one can easily come to the conclusion that the 140 firewalls alone, which would require 24/7 service, are a major cost factor as well as a security risk. Consolidation would lead to huge savings. This may be an exceptional example because DHL/Deutsche Post it not only a worldwide corporation but also large parts of their current infrastructure are the consequence of the worldwide acquisition of logistics companies. But it is nevertheless a good place to look deeper into consolidation processes.

References

[1] Accenture, 2003 study.

[2] *Total Telecom,* July 2003.

[3] "Migrating to Enterprise-Wide Communications: The Branch Office Dilemma," Unified View Consulting, May 2003.

3

Moving from "Bits and Bytes" to "Productivity and Profit"

The future influences the present just as much as the past.
—*Friedrich Nietsche, (1843–1900)*

Whereas every individual has already experienced the changes in the business and private environment that arose from the new networking technology, most corporations have retained the status quo. There has been little, if any, change in the development of corporate strategies for voice and data networks.

Even today, most companies continue to rely on historic approaches for solving both current and future problems. This attitude touches not only on the general questions of how to cope with technological change and how to make the best use of new technology, but also relates to the main issue of information technology and telecommunications strategy in the corporate world.

3.1 Role of the IT Department and the CIO: Different Views and Their Implications

It is impossible to address all facets of the current discussion on the role of information technology and telecommunications within the pages of this book. It is beyond the scope of the intentions and ideas we wish to address and is likely to be unproductive due to the constant changes in professional discussions on this topic.

Nevertheless, it is important to understand the main points of view within this public argument. Therefore, in this chapter, we discuss both organizational and the cost-oriented considerations. Bringing together both of these considerations will enable us to see the entire scenario. The core message emerges: "How can we do more with less?"

3.1.1 Development of IT Spending

National expenditures for the development of information technology create an impression of change in the role of information technology and telecommunications. IT spending follows the economic cycles. The end of the new economy was followed by several years of defensive development, and worldwide IT spending recently made a slight upturn. Several analyst companies confirm this and offer similar descriptions of what is currently happening worldwide. Datamonitor describes IT spending from 2003 to the present as growth oriented, while spending prior to 2002 was mainly driven by cost savings [1]. The situation changed in the third quarter of 2003. Most of the companies worldwide can now be described as growth oriented when investing in information technology and telecommunications.

3.1.2 Cost Issues

A very common view of information technology and telecommunications is that of seeing them simply as expense factors. A common strategy within this attitude is to benchmark spending for IT and TC based on the spending of similar sized companies or companies in the same industry.

The goal is not always the lowest possible percentage value but more advanced cost-driven strategies as these organizations try to mimic the companies they perceive as the best in managing their spending habits. This more concise approach assumes that there is a correlation between information technology spending and company success. Unfortunately, most data disputes this correlation. Recent research from McKinsey indicates no interrelation between information technology spending and company success [2].

Most spending-oriented strategies focus solely on cost savings in isolated areas, such as new voice or data connections, made at certain stages of company development.

In the long run, however, this concept leads to a fragmented infrastructure that tends to be more expensive than a network infrastructure planned in a holistic manner. Despite this, cutting costs is a typical way of doing business in most corporations, not only in times of economic downturn.

Unnecessary infrastructure components purchased in times of economic growth will often be discarded. Infrastructure and service contracts will be changed and updated to reflect the downward price spiral during the recent change from a seller's market to a buyer's market.

Here is a list of the most common cost-cutting ideas:

- Checking and repricing of all current contracts;

- Changing of consulting contracts from a per-day basis to fixed-price terms;
- Concentrating the buying of all hardware, software, and infrastructure services with just one or a limited number of suppliers;
- Checking the necessity of network connections and the possibility of bandwidth reductions on existing connections by testing actual usage;
- Testing all current projects to see if they are as useful as originally thought and if costs and time frames are within the initially assigned values; also includes identifying and eliminating projects that deliver no real added value;
- Adapting existing service-level agreements to actual company needs;
- Differentiation of user needs using a new service model;
- Standardization of hardware, software, infrastructure, and services, which saves on purchase costs through economies of scale and also saves on service and support costs through reduced complexity;
- Examining all contracts for services and other maintenance work to verify their usefulness, and canceling those that are no longer needed;
- Verifying previous make or buy decisions.

Taking a closer look at these procedures, one must admit that even such a simple cost-oriented view of information technology and telecommunications can lead to taking the right steps for consolidation and reinvention of the corporate network. Of course, this will only happen if total cost orientation takes precedence before punctual optimization of cost components.

From a holistic point of view, simple cost optimization often leads to massive problems with a reduction of manpower, information, problem-solving competency, expertise, and, in the long run, motivation and flexibility.

3.1.3 The Value of Information Technology and Telecommunications

Much discussion within large corporations revolves around the value or worthiness of information technology and telecommunications. Such discussion can be dangerous.

The most unrewarding job one can have within these organizations is to be responsible for network security. If the person does well, nothing will happen, which is best for the corporation but he receives no credit; if the system fails, however, he will be held responsible. The value of information security is therefore recognized only after an incident. Typically, these incidents involve data losses and several hours or days of infrastructure nonavailability, resulting in large parts of the workforce being nonproductive for long periods of time. The financial results of such a disaster then, and only then, prove the value of information security.

This problem aside, there are a few different approaches for finding value in IT and TC. These approaches rely on either quantitative or qualitative aspects. A quantitative examination, typically called a *return on investment calculation* or *return on IT*, calculates the financially measurable effects of process improvements or productivity gains. Process improvements can lead to measurable results, such as increased revenue per employee, reduction of distribution costs, reduction of warehousing costs, and an improved ability to deliver. Productivity gains mainly include time savings on designated tasks. In some cases, a reasonable commercial assessment of these effects in concrete numbers remains difficult.

Other positive results are even more difficult to measure or calculate, such as when new technologies act as an enabler of services or tasks that otherwise cannot be done or are necessary for establishing a new business process.

Whatever one does, most of the results remain vulnerable to criticism.

3.1.4 Information Technology and Telecommunications as a Company-Wide Challenge

Another popular view of IT and TC is to view them as company-wide tasks that include different roles. Within these roles, or role models, several tasks must be performed:

- Senior management defines the basic strategy. This strategy includes system architecture, technology platforms, directions for sourcing/procurement, and, of course, the organizational guidelines for the IT and TC department.
- The operational division defines its process needs and capitalizes on the services and infrastructure provided. In a perfect world, innovation would also result from these operational divisions. In reality, this is not always true.
- The IT and TC department is responsible for providing the necessary services in a cost-effective manner. Success, however, is not limited to monetary aspects, and should not be measured solely by financial criteria, that is, staying within the budget, but primarily for compliance with defined service levels.

Figure 3.1 illustrates these tasks.

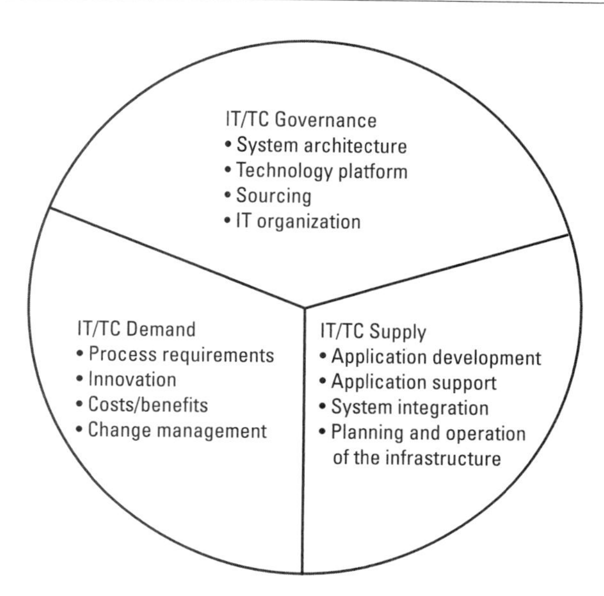

Figure 3.1 IT/TC as a company-wide task. (*After:* Manss & Partner, Hamburg, 2004.)

3.1.5 Information Technology as a Company Service Center

The role of the IT department is often defined as the custodian of the budget and internal supplier of services to operational divisions. This role includes supplier-centric thinking on what services are offered and how they are provided. This view is changing to a more service-oriented understanding of what lies within the responsibility of the IT organization. Operational divisions expect more and more service and customer orientation from their IT department. Sometimes the IT organization suddenly finds itself in direct competition with external suppliers. A lot of companies allow their operational divisions to budget for special tasks and these divisions find themselves in a better position for obtaining what they want from internal or external suppliers. This is referred to as the ROI-driven approach.

The consequences of this change to a more user-centric understanding of IT and TC is a changing role for the IT department that should result in a better understanding between IT and operations and provide them with common values. Within this scenario, a more detailed analysis leads to a separation of the IT and TC activities into three distinct categories:

1. *Run the business.* This category includes all basic services and solutions necessary for doing business, such as messaging and telephony services.

2. *Grow the business.* This category includes all activities supporting the business processes and is helpful in promoting core business growth.
3. *Transform the business.* This category produces growth through reengineering and redesign.

Everything in the "*run the business*" category should be viewed primarily as either minimizing risk or resulting in cost efficiency. Cost efficiency, within this context, is defined as follows:

- Procurement to meet requirements;
- Standardization, as far as it is economically expedient;
- Use of the economies of scale; and, most importantly,
- Testing of possible selective outsourcing.

Everything in the "*Grow the business*" category or the "*Transform the business*" category is treated differently. Project decisions falling within these categories would be made after calculating their value to the corporation, the so-called "return on IT." Before any investment decision is made, a dialogue should take place between the IT department and the company departments who are seen as the potential users of the new system or software. This includes specification of requirements, consulting on business processes, cost estimations, and, eventually, a validation of the business case. The execution of the project, its implementation, and the final evaluation of the outcome will be favorable if there is strong cooperation between the IT department and the operating departments.

The consequence of the approach outlined is a role model, or role understanding, of IT and IT management within the corporation: The IT department changes its role from being a technical supporter to being an enabler of business and a service provider. This implies the need for a detailed understanding of business processes, including business and finance-oriented knowledge, project management abilities, and the supervision of external suppliers.

3.1.6 Technology as a Tool for Doing Business

The rise and fall of the new economy and the accompanying e-business boom brought with it technology, which seemed enough in itself. Projects were begun without foreknowledge of the outcome or possible financial success and often failed. After the subsequent downturn after the end of the new economy, no budgets were available to support projects having unclear financial objectives. This is still true for most corporations today. There is also a tendency to go to the other extreme. The downturn was also used to cut costs and budgets everywhere. More goals need to be accomplished with fewer personnel and smaller budgets: better support of business processes, higher productivity per employee, higher efficiency, and a higher financial benefit. The consequence is that fewer employees have to fulfill more and more and tasks of increasing complexity. Within most corporations, technology is also a tool for achieving differentiation within competitive markets, for example, with new products or services.

To achieve the goals defined by these targets, it is necessary to rely heavily on the interplay of the following components:

- *People* having the skills to define requirements and the willingness and abilities to accept and use the solutions provided;
- *Processes* to be defined and adopted;
- *Technologies* that can be used to support the necessary processes and achieve solutions.

3.1.7 Doing More with Less

Although it is difficult to increase corporate performance under conditions of static or decreasing budgets, this is the reality for the IT and TC departments in most corporations. The Yankee Group, well-known analysts and consultants, produced a widely distributed report in late 2003 entitled *Why Network Managers Can't Sleep at Night* [4]. The report contained the following survey results:

- 51% of all respondents reported pressure from high costs and low budgets in addition to restrictions on adding personnel.
- 46% must change from legacy systems to modern IT architecture, and are concerned about maintaining quality of service and reliability.
- 41% identify network security as their main concern.
- 34% are challenged by the necessity of providing new applications.

Despite all of the discussion about slow economic growth and value-oriented thinking, costs remain the major issue for IT and TC departments within large corporations. Increased spending or budget growth can only be expected for special interest areas as a result of public discussion in both the specialized and general press. (Security is a current example for this causal relationship.)

Changes in external factors and increasing requirements lead to one simple core question: "As the one responsible for IT and telecommunications in my company, how can I do more with less?" This question should be examined in the context of the experience of IT departments that have undergone several cost-saving cycles, are typically understaffed, or have employees with the wrong set of skills.

Increased Complexity

With higher expectations, there is usually increased complexity in almost every aspect of providing IT and TC services within a corporation. The problem is compounded by many additional applications and hardware, as well as a growing numbers of services needed to support the operation. IT departments also encounter difficulties in finding employees with the right skills who are willing to cope with the ever-increasing changes in this area. In small and medium-sized companies in particular, there is a gap between the IT and TC services wanted and needed by the staff, and those that can be provided because of the limited number of personnel available.

With technological advances, an increasing number of small and medium-sized enterprises rely as heavily as large corporations on the network infrastructure and associated services. The only difference between the larger and smaller companies in this regard is the size of the IT staff. Small businesses are typically understaffed and have fewer overall resources.

A medium-sized company acting as a supplier in the automotive industry today typically has production facilities in more than one country. Some of them produce their parts in different time zones and also often work around the clock in three shifts. In this case, the manufacturer will need network and application support around the clock to support these functions. In reality, it is atypical that a medium-sized business has services within their own IT organization available outside the normal 9-to-5 working hours. They simply cannot afford it.

Also, even large corporations often rely heavily on a few employees having the skills and expertise to know and understand every detail of their complex corporate IT and network infrastructure.

Reduction of Market and Product Cycles

Within the last decades we have experienced a massive acceleration of product development. IT/TC product life cycles are getting shorter and shorter. Business processes of all kinds must become faster to adapt to rapidly changing market conditions and ever-changing customer needs.

Within IT/TC the need for upgrades/updates and patches for software that is already in use has become ridiculously fast, partly, but not only, due to almost daily new security threats and the incapability of the software industry to supply an error-free product.

The necessity to implement, adapt, and, if necessary, remove new applications at corporate levels in shorter time frames also leads to increased pressure on the IT department (see Figure 3.2). Proceeding as usual (that is, in the way the IT department used to operate) is often no longer fast enough to meet these requirements.

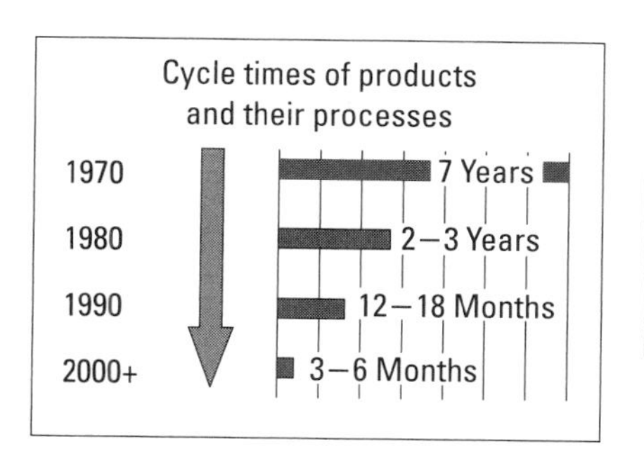

- Processes must be more rapidly adapted to customer needs and markets.
- Conventional IT processes do not meet requirements.
- An increased level of flexibility, implementation, and cycle speed is required.

Figure 3.2 Shortening of market cycles. (*From:* Manss & Partner, Hamburg, 2004.)

Support for Mobile and Home Office Workers

Another big challenge for most corporations is supplying the necessary support to mobile workers and teleworkers (those working from home). In Chapter 1 we discussed the consequences of change for the individual. For corporations, the increasing mobility of the workforce leads to increasing support requirements in the implementation and maintenance of services, for example, services allowing secure remote access to the corporate network and corporate data, the integration of home office working environments, or the provision of messaging services for handheld devices. Besides this, services need to be delivered to workers who move between different corporate branch offices in the course of their work and need access to their e-mail services and telephone direct lines while at different sites.

In the past, only special groups of employees such as the sales force, service technicians, and executives typically needed mobility support services. Mobile services now target not only those employees just mentioned, but also larger parts of the workforce. Figure 3.3 shows different types of employees identified by the INSET Mobile Business Report published in 2004.

Implementation of New Applications

New applications delivering established business processes to mobile devices, as well as new processes that must be supported by software and implemented company-wide, are also responsible for higher complexity and expenditure. If, for example, a new enterprise resource planning (ERP) system must be available company-wide, this has consequences on the bandwidth required within the corporate network and on other parameters of the company infrastructure.

The main focus of newly implemented software nowadays is often voice services and other applications using the data network (voice over IP or VoIP). Most companies are experimenting with this but lack the experience required for the planning and implementation of a frictionless rollout of these systems.

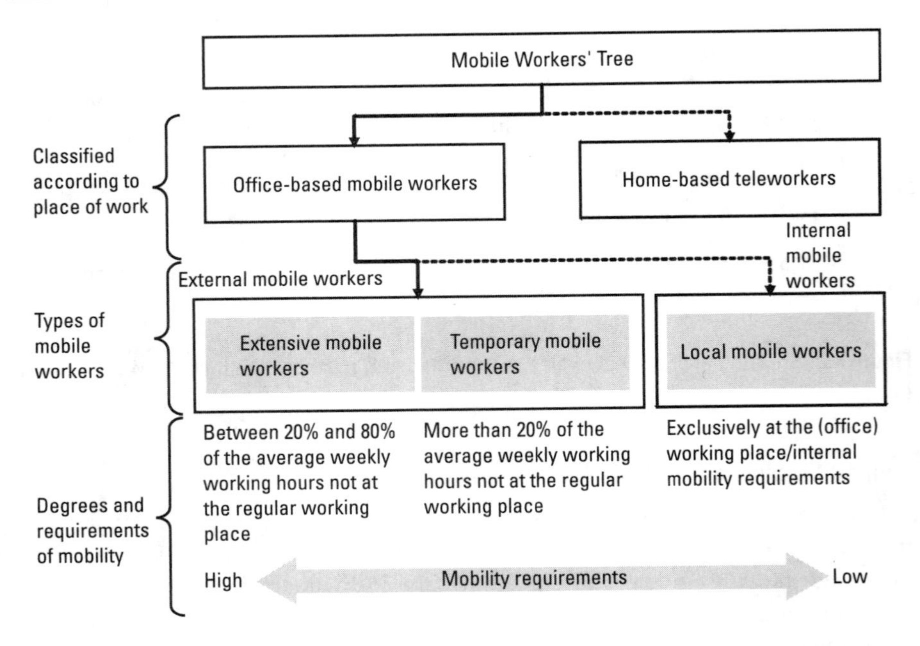

Figure 3.3 Mobile worker typology. (*From:* INSET Research and Advisory, Mobile Business Report 2004.)

The main reason for this problem can be found in the way companies previously supported voice networks: Voice was typically delivered as telephone service over its own network that was supported by a telecommunication supplier as a CENTREX service or as a hardware-based telephone system that was typically only compatible with itself and with the equipment supplied by the particular manufacturer or supplier.

Quality-of-Service and Performance Improvements

Although QoS is not monitored everywhere, it occupies a major place in discussions within IT and TC departments. There is broad agreement on the fact that system breakdowns and long response times in software applications need to be avoided. Against this stands the dominant cost-oriented view of IT and TC services discussed earlier.

Users in corporations of all sizes still often complain about performance or availability problems. These problems very often remain invisible or achieve public acceptance in the IT department only after massive complaints, and consequently too late. In a report published by Forrester Research in 2004 [4], 67% of all respondents (among 430 IT managers questioned) admitted that performance problems in business-critical applications only come to their notice when end users call the help desk. Six percent only discover that there is a problem when top executives complain to them. A total of 41% answered that the company strategy in preservation of application performance is passive.

Securing the Network

Securing the network is a permanent issue in companies of all sizes. With increasing security problems, which are mostly Internet related (viruses, worms, trojans, and other malware), the expenditures for securing against these problems and coping with the consequences of related incidents are rising. With the connection of the corporate network to the Internet (for e-mail, Web access, Web-related services) years ago, the Internet has also become part of corporate WANs, whether one likes it or not, and therefore requires special attention. Attention must also be paid to those people with internal access to software and systems that can use their knowledge to harm the corporation.

Although wisdom about security problems is widely distributed, a lot of companies see no need to invest. A report from Ernst & Young, distributed in August 2003 [5], asked 1,400 IT managers in 66 countries IT security-related questions and came to the following conclusions: 90% of all respondents see IT security as important but only 33% of respondents also see themselves as completely shielded in the case of an attack. Almost the same percentage of respondents admitted that they only vaguely knew if an attack or security incident had occurred and what systems this affected. Only the protection against a breakdown of the business finds above average interest. Sixty-five percent of all respondents see disaster protection and disaster recovery as major topics, while everyday risks are heavily underestimated.

Apart from the general developments in IT security, there is another fatal aspect. Due to cost savings–oriented decisions, there is limited knowledge of the cost and risks of new technology after implementation (consequential costs/consequential risks). Self-managed virtual private networks (VPNs) based on the public Internet, which are often chosen due to cost issues, lead to new security issues. The line cost savings of these heavily promoted low-cost, wide-area connections are often more than offset by additional expenses for firewall management.

Protection Against Corporate Espionage

Speaking of espionage, one usually thinks primarily of the intelligence services of foreign countries and their activities that are sometimes also used with business intentions. But corporations that are rivals in the market also attempt to gain a competitive advantage by systematically collecting and processing competitor data. This process is normally named *competitive intelligence* (CI) or *market intelligence* and is a relatively new but respected discipline as long as it remains within the boundaries of the law. (The author himself teaches university classes in competitive intelligence concepts at the Ansbach University of Applied

Sciences, Bavaria, Germany.) Corporate espionage begins where CI ends. This can occur accidentally, within the heat of the moment, but more often occurs systematically. Unlike CI, nobody wants to talk about corporate espionage, and medium-sized companies in particular are often not aware of these types of espionage risks.

A systematic approach to combating and protecting against this type of hostile behavior of certain competitors, and the intelligence services of other countries, will more and more become an additional issue to be solved by the IT department, at least on a technical level.

Fulfillment of Increased Expectations in Customer and Supplier Relationships

Global competition and the increased speed in the way business must be done in the Internet age has led to the necessity of companies need to immediately provide the following information, among other, to customers and suppliers:

- Order status;
- Delivery status;
- Online ordering capabilities;
- Direct access to selected internal data.

Corporate and end-user customers increasingly expect and demand that precise, detailed information on all data relating to their business transactions be available to them at anytime, not only but often accessible via the Internet. To fulfill these demands, companies must provide the necessary services, despite the cost and budget problems, to avoid losing ground to the competition and, in the long run, losing customers.

Liability Risks

Different legal issues influence the way companies do business and provide the necessary IT services. Especially in medium-sized companies, there is often not much more than a general overview and no deeper knowledge of the risks and potential liability issues related to providing network infrastructure and related services. For example, if the person responsible for the e-mail service within a corporation fails to protect it against malware, and this failure leads to the use of the system for sending viruses or spam, this can make the company using the mail system liable, even if it was operating in good faith.

The laws widely differ from country to country but the conclusion is indeed worldwide: IT-related liability risks are generally on the rise. Insurance companies often deny coverage of IT and TC risks, or imply an obligation so high or costly that, from a business point of view, it is not economically advantageous to buy insurance protection.

Worldwide Production and Distribution

Corporations involved in distributed production that have international branch offices or have established sites around the world typically need an IT department that can provide not only multilingual support but also—and more importantly—longer service hours. This depends, of course, on the centralization of the infrastructure. Loosely connected sites can often rely only on local support within local service hours, but international corporations more often tend to have at least parts of their infrastructure concentrated in one or just a few locations, which automatically leads to alternate shifts or even 24 × 7 needs in the IT and TC service departments. Alternate shifts, instead of the usual 9-to-5 working hours, create a large increase in costs despite the tight budgets.

Organizational Change

With the growing numbers of mergers and acquisitions worldwide, but also with the shortening time frame of reorganization cycles in large corporations everywhere, there is a need to support those changes or the necessary adoption of different infrastructure elements and services. Business units must be integrated or sometimes just cut off, for example, if a part of the business or a subsidiary is sold. Especially in mergers, there are not only technical but also political problems in finding the leading system or system strategy and implementing it; the losing system loses against the other and gives those responsible bad headaches.

There is no indication that this trend will reverse. In a report published by A. T. Kearney in 2003 [6], the data of more than 25,000 companies worldwide were tracked over a 10-year period. The report comes to the conclusion that there are similar development patterns of consolidation within different industries. This leads to an ongoing demand in consolidation for years to come. Only the speed of consolidation is a major differentiator between industries. Everything else is similar.

Even without mergers and acquisitions, and the subsequent consolidation efforts, intraorganizational changes often lead to new challenges for IT and TC. Within just a few years, the European subsidiary of a large American corporation underwent major strategic changes in distribution: from a decentralized/local model (with storehouses throughout Europe) to a centralized approach (only one storehouse for serving the European market) and then back to local storehouses. One doesn't need to examine the details to estimate the workload imposed on the IT department that had to support all of those strategic changes every few months.

Increased Flexibility

Increased flexibility, faster reaction to changes in market and environment conditions, faster adoption of industry trends, and implementation of new technologies—these are the core challenges for the corporate telecommunications infrastructure.

The individual must already rise to these challenges as described in Chapter 1. Mobile and remote working is on the rise and occupies larger and larger parts of our working hours (see Figure 3.4). This occurs partly as a consequence of technological progress but also as a consequence of the increased pressure companies place on their workforce.

So-called *nonterritorial offices*, where employees just use whichever empty desk is available, in contrast to the traditional method of every employee having his or her own desk regardless of whether they are actually using it, are becoming more common, as are the trends toward working on the move or working from home or at a customer's site. This organizational model is often established to save office space and is therefore often found in innovative industries such as advertising or consulting as well as in corporations located in areas where real estate and office rentals are expensive. This development is also supported by a change in employment from regular contracts to more and more part-time and secondary work.

Even those who have regular contracts are also increasingly affected by changes. Relocation within corporations has increased substantially during the past few decades. If teams work together only for a particular project, then moving from one office or office building to another occurs routinely several times a year. The costs incurred with every move, whether it is only for a few months or the rest of an employee's working life, are very often the same. Depending on

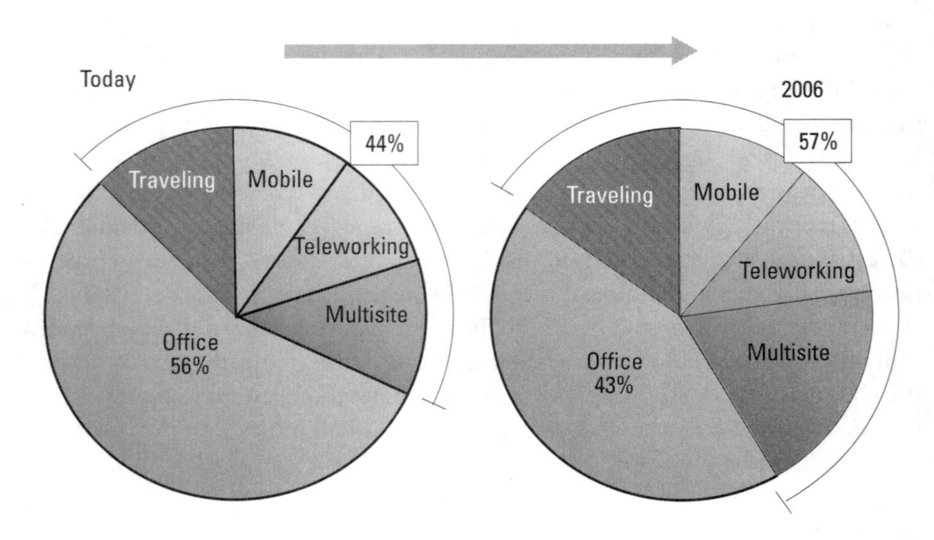

Figure 3.4 Flexibilization of the working environment. (*After:* ITAC, Executive Systems Research Center, Siemens.)

the study you refer to, the costs can be as high as $5,000 per relocation of a single employee. Bringing these costs down not only affects the total operational expenses, but can also make the company more flexible.

Another problem with the infrastructure that opposes the requirement for flexibility is the typically high rigidity of everything that is infrastructure related. Different studies and reports come to the same conclusion almost everywhere in the developed world. Two-thirds or more of the total costs of information technology and telecommunications within corporations are typically fixed costs. This means that there is little room for change and also means that, if the economic situation of the corporation is worsening, it is not possible to reduce the costs of the technological infrastructure to the same degree as the decline in revenue. The corporation is normally stuck with almost all of the costs associated with the planned basis. For example, if you have 2,000 telephone extensions (one for every employee), and lay off 500 employees, you will still have to pay the base costs of 2,000 extensions, although you will probably save on the cost of calls. (This wouldn't be true, of course, if your telephone service is on a flat-fee basis.) Generally speaking, the same thing happens with data networks (where you often have flat fees, so that bringing down bandwidth is often the only way to save costs).

On the other hand, if business goes well, you will probably get into trouble if you need 200 more extensions but your switchboard can only support the amount you already use. You must then move to new, bigger hardware. Executives often push a claim for switching from fixed to variable costing in information technology and telecommunications in order to better support changes in company size and revenues, whether up or down.

Directly following the discussion on flexibility is the discussion of the increasing speed with which we all live. During the few years of the so-called new economy, speed was everything. One recalls the saying: "Internet years are dog years." This did not end after the Internet hype. At a corporate level it has even increased. One must only look at e-mail and how it is used today. People tend to answer messages faster and faster, at least those that seem important to them, and carry around little devices enabling them to do this anytime and anywhere. At the same time, the amount of information that, for example, an executive director must read has also dramatically increased as well as the portion of the workload that is done outside the boundaries of the corporation.

More and more companies, mostly service or development-oriented firms, make use of the so-called "follow the Sun" principle. Teams in different time zones around the world work on the same topic but move from one team to the other, in a shift-like manner, around the globe throughout the day. The consequence: reduced development and production cycles lead to even quicker movement.

3.2 Information Technology and Corporate Networks: Complexity and Commodity

Since the 1990s, the diversity of software systems and available services has increased dramatically within the corporation. E-mail, e-business applications, extranets, and intranets have led to more internal communication as well as more interaction with suppliers and customers. Even with the decline of the e-business hype, core innovations of the Internet age such as VPNs and VoIP (see Chapter 4), as well as intraorganizational discussions, are in the center of public attention and are adopted sooner or later.

The consequence is a highly complex, potentially insecure, more or less open structure based on networking components, leased lines, and systems and applications of different kinds on different platforms that must work together in all details to provide the best possible service quality for the end user. Every minute of downtime leads to financial losses due to lost revenue and lost paid working hours. This infrastructure is expected to function in the same way as do water and electric power supplies: "always on," 100% availability, no downtime, and consistent quality. This often used analogy between network infrastructure and services and water/electrical power includes a tragic component since it also implies that the team responsible for the corporate network and services is not as good as the electricity or water supplier (we won't discuss California here!).

Instead of looking at the infrastructure from a strategic view, most companies still rely on a "bottom-up" approach. Every component of the infrastructure, every server, every router, every switch, every firewall, and every application or storage system is regarded and administrated as a single item. Every item has its own operating status that is individually measured, logged, and analyzed. To analyze this information, there are a large number of software products and tools for network and application management that help to integrate these single tasks into a general view on a centralized console.

The problem with this approach is the missing link to the business view. The aggregated information sources do not give us any information on what topic is really important for concrete business functions or which element should be given priority when troubleshooting. There also is no overview on interdependencies between different areas. The network manager is drowned under a multitude of messages without a link to business aspects, while the user suffers other problems, for example, problems with application performance and varying response times and has no chance to realize that his experiences are a symptom of a structural problem arising somewhere else within the corporate infrastructure. His fault report often goes unanswered, getting lost in the structures and procedures of the corporate help desk, especially if no rules of deduction from similar problems are stored in the help desk database or in the brains of the support personnel.

Under the dictum of "doing more with less" described earlier, a holistic view of the corporate voice and data infrastructure is indispensable. Only this can solve problems resulting from the interaction of the individual components.

A top-down approach is also useful and necessary when using managed services. *Managed services* refers to the situation in which an external service provider is responsible for operating a company's infrastructure. A major precondition for this kind of service is a common understanding of the business processes of the customer and not only of the infrastructure itself.

This change in thinking does not occur overnight. The best way to begin is as part of the normal day-to-day business, such as the necessity to integrate another company as a consequence of a merger or acquisition, or a decision to outsource central services to an external provider.

3.3 The Changing Role of the CIO

Since computer technology and data networks found their way into the hearts of all company functions, from accounting to logistics, and PCs found their way onto every employee's desk, the importance of the person who is in charge of information technology and networking has undoubtedly greatly increased. However, a strong tension between general management and IT management can be felt in most corporations. This is the result of the implication of a relationship that is often felt as a dependency on IT and the special knowledge of IT management. The rise of the IT manager from a simple functionary to the chief information officer (CIO) as a top executive, sometimes an actual member of the board of directors, underlines the strategic importance of IT for the corporation.

The change from an operations manager to a top executive position as an IT strategist is significant. An understanding of technology and its possible impact was, and is, largely seen as a major tool in the fight for markets and market share. But this concept has come under attack within the last few years after the downturn of the new economy. Dozens of business and management books revolve around this topic. The discussion culminated in 2004 in Nicholas Carr's *Does It Matter? Information Technology and the Corrosion of Competitive Advantage* [7]. Carr does not believe in the strategic value of IT. He sees faster adoption of new developments, increasing standardization, and more widespread knowledge of technology as signs that in the future IT will no longer be of strategic value. In his opinion, IT and telecommunications are irrelevant for the success of a business. Even though Carr came under heavy attack after publishing this, the worldwide success of his thesis is a sign that it did indeed hit a sore spot.

Leaving this discussion behind and looking at the more established ideas of common management theoreticians such as Charles Handy, Peter Keen, or John Henderson, one finds three fundamental management roles for IT:

- The *classic role* of IT management is to be responsible for the operability of the network, hardware, and software. Nowadays this also includes taking future business requirements into account and incorporation of new technologies and developments.
- The *business role* requires taking responsibility for the creation of value within the corporation and adds a more active component.
- *IT governance* is an executive task. Steering and surveillance follows business goals.

How do these aspects now come together? In which direction will IT and TC management evolve?

The current discussion about the role of IT and TC is evidence that this development is at a crossroads. For the mid- or long-term future, there are probably only two possible directions: The person responsible for IT and TC will realign herself and evolve into a strategic spin doctor and helmsman at an executive level, but will lose operational functionality. She may also possibly lose one or many of her subordinates as a result of her own outsourcing decisions. She may even find herself in a new role as a coordinator of external service providers.

The other direction is uninviting: If the idea that "IT doesn't matter!" has gained ground within the executive board, the IT/TC manager is possibly on his way back to being a simple "operations guy" and probably won't be asked if outsourcing decisions on are on the table. In this scenario he will possibly no longer be needed in the long run.

References

[1] Datamonitor, Analyst interview, 2004.

[2] McKinsey Presentation at T-Systems NWS Roadshow, Germany, Nov. 2004.

[3] "Why Network Managers Can't Sleep at Night," Yankee Group, 2003.

[4] Forrester Research, February 2004.

[5] Global Information security Survey, Ernst & Young, August 2003.

[6] Kearney, A. T., "Merger Endgames: Industry Consolidation and Long-Term Strategy," ITAC, Executive Systems Research Center, 2003.

[7] Carr, N. G., *Does It Matter? Information Technology and the Corrosion of Competitive Advantage*, Boston: Harvard Business School Press, 2004.

4

New Services: The Value of Benefits for Companies

> People are very open-minded about new things—as long as they're exactly like the old ones.
>
> —*Charles F. Kettering (1876–1958)*

As discussed in Chapter 1, the boundaries of the company as we have seen in the past have been broken down. The new understanding of company operation beyond its spatiotemporal borders requires the support of secure and flexible network connections. Secure and flexible connections are needed between company locations as well as between the company as an entity and mobile workers or teleworkers. Typical WANs are based on leased lines between the company headquarters and branch offices (or at least the larger branch offices). Smaller branch offices and teleworkers are often connected in a much simpler way, via dial-up or by using the public Internet. If you step back and examine this corporate network en bloc, the shape is similar to a star with lines extending from the center (headquarters) to the endpoints (branch offices and teleworkers). This star shaped network structure is also called a "hub-and-spoke" architecture.

As long as most services were centralized, this was more or less the ideal architectural model for a company WAN. With more and more decentralization, and more and more cooperation between branch offices, and between teleworkers and branches, the necessity for a more mesh-like structure arose. Virtual private networks are a possible solution to this dilemma. VPN basics, as well as VPN applications for business, are the main focal points of this chapter.

The technological innovations of the last decades not only affect VPNs but also have a fundamental effect that we call *convergence*. Voice and data traffic converge on the same network: Voice over IP, the transmission of voice on data networks, plays the core role of a truly disruptive technology. The same is true of what is sometimes called *rich media.* In rich media, in addition to voice traffic, video is also integrated into the data network.

An important precondition for bringing this voice-data convergence to a useful application level within the company is quality of service. Only with a defined quality, for example, that provided by multiprotocol label switching (MPLS) technology, can the positive aspects of voice-data convergence show their full advantages.

4.1 Virtual Private Networks

VPNs are very important as a basis for providing universal availability as defined in Section 1.1.2. A VPN connects two computers or two private networks with each other and uses another network for the intermediate connection. The intermediate network is often a provider network or a public network, such as the Internet.

A VPN consists of two or more VPN endpoints. If the VPN consists of only two components, one speaks of a *point-to-point* connection. If there are more than two components within the structure of the VPN, one speaks of a *multipoint* service.

This sounds like a weak description but it is the common basis for all definitions of a VPN service.

4.1.1 Not All VPNs Are the Same

The IT and TC industries are driven by buzzwords and acronyms and most players try to adapt their offerings to a current trend or a buzzword that is in vogue. One can see the consequences of this when it comes to VPNs. A comparison of press releases and product sheets for half a dozen vendors easily reveals half a dozen or more different definitions or understandings of the term VPN.

Where are the similarities and differences of current VPN definitions? A core aspect of every VPN is the logical network that is provided independently of the underlying topology of the physical network. To get closer to the core of the term, one can separate it on the basis of the words in the acronym: *virtual, private,* and *network:*

- The term *virtual* refers to a group of computer systems distributed in space (at more than one location) in which every system can interact with each other and which can be managed as a single network. The possibilities and opportunities lying within a LAN are extended without the need to consider geographical boundaries.

- The term *private* must be used with more differentiation. Some vendors define *private* as "nonpublic" and, therefore, from a technical point of view, as "closed user groups." Within the scope provided by this definition, the acronym VPN can be used for a wide area of carrier services, from frame relay to ATM- and MPLS-based networks. Another interpretation of the term *private* sees security as a core aspect. Authentication, confidentiality, and information integrity between the users and computers are the main considerations. The minimum requirements for a VPN, therefore, are that no one should be able to see the data traffic of a VPN or should be able access it unless he or she is an authorized user. In an ideal world, outsiders (nonauthorized VPN users) should not even know of the existence of another VPN on the same network. In practice, a VPN based on the same non-QoS-enabled infrastructure together with other VPNs (e.g., the public Internet) can compromise other VPNs due to a high traffic load.
- The term *network* does not need a detailed explanation in the context of this book.

4.1.2 Use of VPNs

The language used by the service providers gives a very broad spectrum of what can be called a VPN. Even so, the most important applications of VPNs are described here.

LAN–LAN Interlinking

One of the most important applications for a VPN is LAN–LAN interlinking. Two or more local networks are connected with each other through another network. From the point of view of the user, it is just one network. All necessary network services can be used from any point within the virtual network structure.

Besides this network–network coupling within a VPN, single PCs from different networks can also be connected in a logical network that is distinct from the other networks. This feature can be used to separate company departments from the rest of the network. Within banks this may be necessary in order to support the concept of "Chinese walls," which separate departments from other departments to make sure that no conflict of interest arises. Within a network structure not only an end user but also a whole company can belong to one or more VPNs.

Remote Access

Remote access is another important application for VPNs. Home office users (or teleworkers) and mobile employees (mostly but not exclusively from sales, service, or management) use VPNs via the public Internet as a transport medium for connecting to the corporate intranet and to access all the important company resources and applications.

Within the last few years, remote-access VPNs have widely replaced remote dial-in systems. The necessity of using a modem (limited to 56.6 Kbps) to call the corporate network via a company-owned or provider-owned dial-in server via the telephone network (sometimes on a long-distance call or international call basis) has been replaced by the use of locally available Internet access (often a broadband connection). DSL connections or cable modem connections are very often used for teleworking, whereas mobile workers increasingly use public WLAN access points and mobile data services (e.g., GPRS or 3G data). All of these replace the historic modem connections more and more. Only in really remote places are modems still used.

While the advantages of remote access are clear, there are of course disadvantages. The main point here is security. A VPN connection via a public network is (or at least should be) secured against attackers by being encrypted. A weak point is very often the notebook or desktop computers of the employees at the endpoints of the connection (mobile worker or teleworker). If an attacker gains control over such a PC system, the door to the corporate network (or at least a major part of it) stands wide open. Viruses, Internet worms, and other malware, which can be loaded on the computer by visiting manipulated Web sites without the user's knowledge, or spyware, which rides piggyback on legitimate software or comes hidden within software backed by license contracts of several thousand words that nobody reads in detail, allow control of the employee PC and can use the secured connection to reach into the heart of the corporate network. Almost every month a new study comes out underlining the threats to the corporate network from the large number of compromised or easy-to-compromise client PCs outside the core infrastructure.

Security software such as virus scanners or personal firewalls on the remote PCs are no guarantee for security, since the users very often disable these functions in part or in whole to get their favorite programs (such as file sharing) up and running or (from their point of view) free the PC from unnecessary tasks and speed it up (by stopping time-consuming search processes).

One step toward more security when remotely accessing the corporate network is increased configuration control using automatic (forced) updates for the client PCs at the network borders. As long as hardware and software are provided by the company, another simple but effective measure would be banning the use of private software on the system. The employee will see this as restrictive, but it effectively decreases the security risks of remote access.

Extranets

Remote-access VPNs as well as VPNs for LAN–LAN interlinking can also be used to connect business partners and customers to allow them access to a corporate extranet or special business-to-business (B2B) services. Especially within these implementations of VPNs, the major advantage of flexibility (simple rollout and configuration changes) comes into play. Within an extranet the same applies as mentioned earlier for a VPN: A company or an end user can be a member or a part of several VPNs.

4.1.3 Acceptance of VPNs

What benefits do companies see in VPNs? What is then relevant for the acceptance of VPNs? A 2004 study by the Yankee Group [1], an analyst and consulting company, showed what matters most by presenting this question to a number of large and multinational corporations (percentages are in brackets; more than one answer was allowed):

- Cost savings (46%);
- Simplified administration (40%);
- Integration of new/additional branch offices (32%);
- Increased bandwidth needs (29%);
- New applications (27%);
- Voice-data convergence (27%);
- Increasing need to connect the branch offices with each other (and not only with the headquarters) (16%).

More than 87% of all large and multinational corporations therefore already had a VPN of some kind in place in 2004, compared with 22% of small and medium-sized businesses. One can conclude that the adoption rate depends on the company size.

Within the small and medium-sized companies, the answers to the general VPN acceptance question produced different figures:

- Increased bandwidth needs (30%);
- Application performance (29%);
- Broadband access (27%);
- Connection of remote users (26%);
- Support of mobile users (26%);
- Better customer service (25%).

Although the use of VPNs is already relatively widespread within corporations, there will be changes within the VPN market, especially given the wide definition of what a VPN is and what it can do.

Most currently implemented VPNs, for example, are self-managed or are based on the public Internet and therefore are a potential place to start consolidation. This is especially important since almost none of the current implementations support the QoS necessary for services such as VoIP. The advent of VoIP very often leads to a new approach to the VPN theme.

4.1.4 Advantages of VPNs

VPNs have a number of benefits, especially when compared with corporate WANs based on leased lines:

- Lower cost of ownership;
- Lower capital expenditure;
- Simple scalability ("on demand");
- Easy and fast adaptation to changing environmental conditions.

Additional benefits come from provider-based VPNs:

- Defined QoS possible;
- High availability;
- High security level/risk avoidance;
- Higher flexibility than leased lines when it comes to bandwidth, redundancy, QoS, SLA, and transparency.

To sum this up, VPNs can be seen as key components for convergence, not only for voice-data integration but also for simplification and integration of once disparate network platforms such as frame relay and ATM.

4.1.5 Internet VPNs

A special type of VPN that has become popular, mostly as a low-cost solution, is VPN over the Internet. Internet VPNs use the public Internet as a transport network. The connections between the VPN endpoints are technically and organizationally based on so-called tunnels/secure tunnels. The management of the VPN network is done by the company itself; the provider network has no knowledge of the existence of the VPN and simply forwards the IP packets to their destination. Another term for this kind of VPN is *company-based VPN*.

Although some service providers also offer support and service for Internet-based VPNs, it still remains a customer-oriented application and is different—from an architectural point of view—than the provider-based models described later.

At first sight, Internet-based VPNs are cost effective and can be rapidly implemented without additional coordination and other measures. An Internet VPN can also be implemented over several provider networks, which makes international connections easy to deploy. But the disadvantages must not be ignored:

- The quality of the network as a whole is not predictable and measurable.
- There is no protection against breakdown and downtimes are not predictable.
- System security depends on the quality of the encryption algorithm used. With "do-it-yourself" VPNs in particular, one must have doubts.
- Administration becomes increasingly difficult as the number of endpoints increases. Scalability is limited and transparency is lost.
- An element of security risk remains, no matter what measures are taken, due to the open nature of the underlying network

4.1.6 Secure Socket Layer VPNs

Secure Socket Layer (SSL) VPNs are a subtype of Internet-based VPNs. SSL is a base technology for secure Internet communications and is implemented in every standard browser. SSL VPNs are therefore easy to use for extranets and remote-access applications since they are easy to administer and do not require a separate client or device at the user's site (since SSL is integrated in the browser). Besides access to Web-based applications, there is also the possibility of using other applications such as Outlook or Notes with the help of additional browser extensions such as Java or Active-X controls.

Security for SSL VPNs is a major concern since SSL-based traffic cannot be checked by firewall systems. SSL also leads to higher resource consumption on server systems.

With SSL VPNs the company is responsible for system administration and support.

4.1.7 Provider-Based VPNs

The major alternative solution for an Internet-based VPN with its own infrastructure built on top of dedicated leased lines is VPNs over provider networks or provider-based VPNs. Provisioning and operation are done by a network provider using its own network infrastructure and not using the public Internet. An ideal service includes a complete "end-to-end" service. With additional mechanisms, quality guarantees are also possible—a major difference to the best effort approach of Internet-based VPNs. The benefits for customers are as follows:

- Lower costs. Normally VPNs over provider networks have lower costs than does the option of building your own WAN on top of dedicated leased lines
- The company receives a complete managed structure and needs few or none of its own personnel to handle it.
- Availability is high, especially when looking at a cost/availability relationship.
- A provider-based VPN is more flexible than the usage of leased lines and easily scalable.
- Provisioning of a defined QoS is possible.
- There are clearly defined service-level agreements (SLAs); also the response to errors or problems that may occur can be differentiated to meet cost demands.
- The infrastructure is totally transparent from a user's point of view.
- Data security, privacy, and data integrity are better than in Internet-based VPNs.
- A provider-based VPN is relatively independent of technological progress. (No CPE can become outdated because the VPN usage time frame and new developments are supported by the VPN provider.)

The network provider/carrier also benefits from VPNs:

- There are synergies in the provider network compared to selling dedicated leased lines.
- Based on new fiber-optic technologies, VPNs can be provided at relatively low cost.
- Administrative work is reduced.
- High flexibility is possible via the configuration of parameters: technology, bandwidth, redundancy, QoS, SLA, transparency, and access.

The positive effects described for providers also lead in the long run to benefits for the VPN subscriber when it comes to cost and QoS.

4.1.8 VPN Taxonomy

It is not easy to find the way through the jungle of all technologies subsumed under the term *VPN.* Besides the distinction between provider-based and Internet-based VPNs, there are other distinctions. Consider, for example, the difference between so-called layer 2 VPNs and layer 3 VPNs. This distinction has consequences on the usability of legacy data protocols such as IPX, SNA, or Appletalk within the VPN, which are only usable with layer 2 VPNs, whereas layer 3 VPNs only support IP.

Other distinctions can be made by looking at the underlying provider network (standard IP network, MPLS network, ATM, frame relay, SONET/SDH, telephone network), the tunneling protocol used (IPsec tunnel, L2TP, PPTP, MPLS-LSP, ATM-VP/VC, frame relay VC, SONET/SDH VT, PPT/dial-up), and the type of connection (point-to-point or multipoint). It will not help the aim of this book to go into greater detail here, so we will leave it at this short list.

4.1.9 VPN Outsourcing

The decision between the company setting up a VPN by itself based on, for example, already available public Internet access or the company outsourcing the VPN deployment to a provider depends not only on cost but largely on corporate structures:

For integrating mobile workers and teleworkers into the corporate network, and giving them access to e-mail and company data a simple, self-deployed installation that makes use of the public Internet and maybe uses a browser only will probably suffice.

To connect several branch offices and include business-critical applications (such as centralized ERP systems) that must be available throughout the network at a defined level of quality and security, one usually must outsource the provisioning and operations.

More and more VPNs are outsourced to external service providers. Within this group there are specialized service providers, local ISPs, but most of all large telecommunications carriers and their solution houses. In the outsourcing scenario described here, the service provider is responsible for the setup, configuration, operation, and permanent monitoring of the VPN. Necessary devices are located within the provider's network as well as at customer locations.

4.2 Voice over IP

The *Telecom Dictionary* [2] defines voice over IP as follows: "A process of sending voice telephone signals over the Internet or other data network. If the telephone signal is in analog form (voice or fax) the signal is first converted to a digital form. Packet routing information is then added to the digital voice signal so it can be routed through the Internet or data network." Voice transmission over IP networks promises a revolution for enterprises.

No analysis company and no major consulting house these days are without a study on the development of VoIP. All show increasing numbers of enterprises using VoIP. For example, the Californian market research company Radicati Group expects that by 2008 about 44 % of all companies will be using VoIP [3]. Radicati's study also describes the main reasons for doing so: "Cost savings" and "integration of new applications" are the main driving influences for VoIP deployment.

Although anyone who has been in this industry for a while will be skeptical of any hard numbers or expected percentage values presented within these research papers, the key message is the same almost everywhere: Voice transmission over IP networks is the next big thing.

But the already-implemented technologies are expected to persist during this change from traditional voice transmission to VoIP. More than 100 years of development builds the basis for traditional phone services. Although there was not much innovation compared to the IT industry, the last 100 years have created expectations based on experience with the traditional services that must be fulfilled by any new technology intending to replace it:

- High voice quality. (At least in fixed-line telephony, mobile phones have already conditioned the user to accept lower quality.)
- Almost 100% reliability. No phone user will accept normal IT downtimes when using the telephone network. As soon as the handset is lifted from the hook, everyone expects a dial tone as a sign that the phone service is immediately working.

The barrier of expectancy is high for any new technology intending to replace the classical phone service (within companies but also within private households). To fully understand the impact of this development, it is useful to take at more detailed look at the basics.

4.2.1 Network and Connection Types

Today's networks are based on two different base technologies that connect end devices (or terminals) with each other: (1) line switching (traditional, typical for voice networks) and (2) packet switching (typical for data networks). This distinction has survived to the present day. In most companies these networks exist side by side as the company phone network and the company data network (LAN).

When using voice networks for data transfer (e.g., for Internet dial-up via the telephone network using a modem, or for connecting remote company branches via dial-up connections), one easily discovers that these line-switching networks are not ideally suited for data transmission.

Normal data transmission patterns include time periods with high activity levels but also time periods with no activity at all. Periods with high activity usually form a relatively low share of the total transmission time. Although this may seem strange at first, one simple example shows why: "Surfing the Internet" includes activities such as page requests (the user types in the URL of the Web site), transmission of the Web page content from the server to the user browser, viewing and reading of the content (during which no further transmission occurs), then page requests again, and so forth. Other server-based applications have a similar usage pattern, with phases of high activity and phases of low to zero activity.

Most server-based applications need a permanent ("always on") connection between the server and the client on one hand, but also have long phases of little or no activity on the other, which leads to an overall low usage and high costs when using line switching (since the line must be reserved by this connection during the whole session regardless of whether or not transmission occurs).

The main reasons for the cost issue are the tariff models used for line switching. One normally must pay for the amount of time the line is "rented." Flat-rate pricing (in which price is independent of usage) was available in certain dial-up scenarios, but because line resources are tight, this concept was never really accepted as expected. The consequence is simple: Packet switching is the first choice for data transmission. Various transmissions can take place at the same time, because all data are cut in packets that independently find their way over the network to the destination points (almost) at the same time.

The dominant technology within packet switching is IP. IP allows for the simple transmission of data within packet-switched networks and can—with the help of additional protocols such as the Transmission Control Protocol (TCP)—support the additional requirements that are necessary for use within the corporate environment, for example, ensuring the correctness and completeness of transmission. TCP and IP are very often mentioned together. TCP/IP is the de facto standard for data networks nowadays.

4.2.2 VoIP and Convergence

During the last few years, *convergence* was and is one of the most popular terms when discussing telecommunications. The integration of voice and data on the same network is a major factor in all convergence-related talk, and lots of papers and books have been written on this topic. One also speaks here of *converging networks.* This usually refers to IP packet switching.

The term voice over IP already includes this technological advancement. VoIP means transmitting voice calls via packet-switched IP networks instead of using the line-switched networks usually used for voice transmission. Another description would be that VoIP includes different concepts for real-time voice transmission over IP networks. Another term used in this context is *IP telephony*. Some vendors also use the term *broadband telephony*.

To be strict in using the definitions, one would have to admit that IP telephony and VoIP are not exactly synonymous. IP telephony uses VoIP, but it is a software application that is available for different platforms from different vendors. IP telephony includes VoIP applications within corporate networks as well as telephony via the Internet (Internet telephony). The following modes of operation are possible:

- Voice call from PC to PC;
- Voice call from PC to telephone;
- Voice call from telephone to PC;
- Voice call from telephone to telephone.

In the mid-1990s, making phone calls over the Internet was already being discussed in the press. Back then there was a lack of acceptance among early users due to long delays and inconsistent quality. Typical of this first phase were calls from PC to PC (via microphone and speaker or headset).

The main advantage back then—low- or zero-cost telephone calls—was not strong enough to compensate for the disadvantages in the eyes of the users. Therefore, only small numbers of users adopted IP telephony in the early stages. Almost 10 years later, with the availability or more bandwidth and better network quality for the individual user, this has begun to change. Note that if the technical problems are solved, VoIP can do much more than just provide us with cheap phone calls, as discussed next.

4.2.3 Reasons for Using IP Telephony

Back in the 1990s, telecommunication carriers experienced historic changes in their businesses. At the end of the 1990s, data transmissions overtook voice transmissions. Several estimates expect a huge difference between data and voice traffic in the future, within carrier networks up to a 4:1 data:voice relationship. Voice transmission was once the core business of carriers but is now becoming more and more of a side product. Large carriers worldwide have already reacted to this change and use IP in their backbones, even for voice transmissions.

Another factor for change is the growing user interest in value-added services. From a carrier's point of view, this is one of the big chances for compensating for losses in their line businesses, sinking prices, and the shift away from what once was their cash cow: voice transmission. These considerations, mostly driven by carriers and network equipment vendors, sometimes hinder the wishes and requests of the customers. One must now ask: If PSTN works so well and has been established for decades in corporate environments, why should we now implement VoIP, which is still in its infancy, as a technology?" One usually receives answers in one or more of the following categories:

- "VoIP saves money!"
- "VoIP is inevitable!"
- "VoIP delivers possibilities for new applications."

VoIP Saves Money

Early applications of Internet telephony focused only on its money-saving aspect. The goal is to save money on local as well as long-distance and international calls. Call-dependent costs are zero, provided that hardware, software, and fixed-price Internet access is available and paid for. This extreme scenario normally only comes true for real Internet telephony, where call origin and destination are within the boundaries of the Internet. All other scenarios route the call over the Internet as long as it is possible to the most cost-effective PSTN gateway. The savings achievable by doing this have declined during the last few years, due to a decline in prices for both long-distance and international calls. Besides local calls that were available within call packages as free, nowadays several carriers worldwide also include long-distance calling in a flat-fee pricing scheme. So call cost is possibly no longer the main reason for migrating to VoIP.

Another cost-saving potential of VoIP comes into view when examining corporate networks. Instead of provisioning and operating both a voice and a data network next to each other, an integrated network can save money. The infrastructure costs, as well as the operational costs of a single integrated network, are lower than with two separate networks (see Figure 4.1). This is not only true for carriers (as mentioned earlier) but also for companies of all sizes.

This simplification can also lead to later cost savings when changes in the (now unified) infrastructure become necessary. Network extensions, relocation of devices or users within the infrastructure, and other changes can be made faster and at lower cost. Enhanced flexibility also goes with this. When reading current publications on VoIP, one can easily discover that the most common factor within all these is reduced relocation costs.

VoIP Is Inevitable

Nobody will deny that IP is the de facto standard for data transmission and will be the basis of future voice transmissions. Even the vendors of traditional

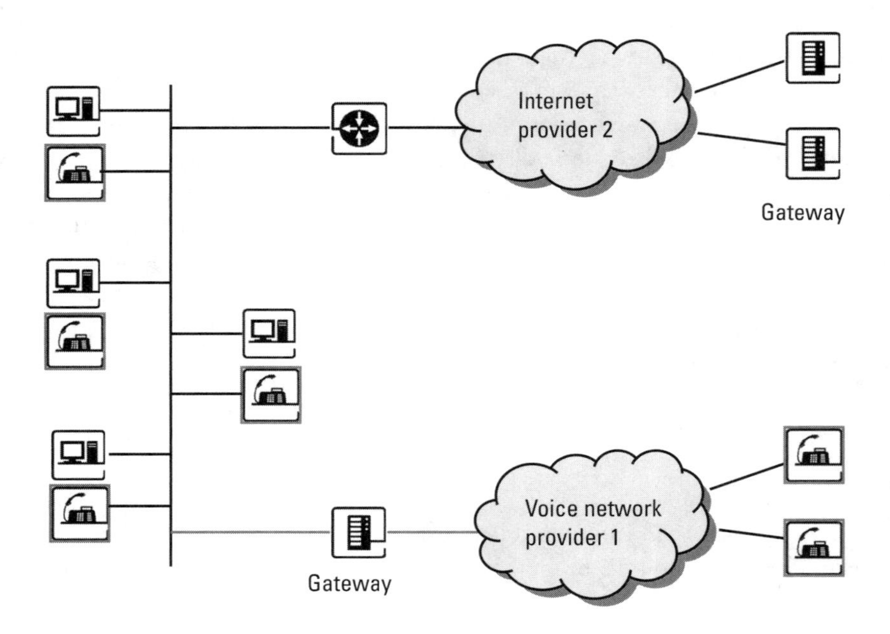

Figure 4.1 Convergent campus network using VoIP for voice transmission.

PABX/PBX systems are migrating their technology platforms more or less to IP. It is improbable that there will even be a next generation of telephone switches, since most developer resources are clearly focused on VoIP.

In summary, one can say that, in the long run, a variation of VoIP will become the only option for corporate voice telephony. VoIP will be, or is already becoming, the next voice transmission technology for home users. Telecommunication incumbents all over the world have already accepted this fight. While still fighting territory losses in the traditional phone service, they are sharpening their offers for VoIP services. This development is partly driven by new players who can start their business without the burdens carried by the traditional players and who are already gaining significant market shares in certain widely deregulated markets.

VoIP Delivers Possibilities for New Applications

IP telephony applications and devices offer a range of new functionality in conjunction with the right software products (see Figure 4.2). The features include unified messaging with media conversion (e.g., saving incoming voice calls as audio files attached to e-mail messages), number portability within the company's own network, multiple-phone ringing (in case of an incoming phone call, several phones ring in parallel), voice and videoconferencing, instant messaging, and workgroup functions such as whiteboards.

Topology	Calls and call management: Voice and unified messaging system	Voice transmission in the WAN	Conventional telephone network
One isolated site	Local	Does not occur	All external calls
Various independent sites	At every site	Does not occur	All external calls
Several sites distributed call processing	In the main headquarters	Yes	All "off-net" calls
Several sites central call processing	At every site	Yes	All "off-net" calls

Figure 4.2 Applications of VoIP.

From a service provider's point of view, the possibility of additional revenue is the main impetus for offering new services. Integration into business applications, for example, via voice portals or IP call centers can be even more useful.

VoIP Advantages Arising Later During Use

Not all the advantages of VoIP implementations are immediately visible on deployment. The advantages visible immediately after implementation of a VoIP solution include the following:

- Easier maintenance and administration due to simplification of the infrastructure on a convergent network;
- Lower expenditures for cabling (in greenfield scenarios);
- Increased functionality for existing LANs;
- Easier disaster recovery and avoidance since components can be easily distributed and are therefore more failure resistant.

Advantages visible later during the course of use are as follows:

- Simple and cost-effective relocation, additions, and configuration changes;
- Access to voice functions from outside the company, which leads to increased productivity for teleworkers and mobile workers;

- Simplified Web-based configuration for telephone hardware and software within the entire company's LAN and WAN systems;
- Better reporting and cost management;
- Productivity gains through additional available applications (unified messaging, voice integration, audio and videoconferencing).

Since most development within company telephony solutions goes into VoIP, these lists will become longer in the future. Traditional PABX/PBX systems will soon be a thing of the past.

4.2.4 Disadvantages of IP Telephony

There are also some reasons to avoid the implementation and usage of VoIP. Some companies are possibly quite happy with the traditional telephony applications they already use and a constant linear business model means little or no need for changes. In this case, the pro-VoIP reasons listed earlier do not fall on fertile ground. It is then not easy to build a functioning, defendable business case for VoIP. The following points speak against further acceptance of VoIP:

- The tariffs for long-distance and international calls have already fallen and are continuing to fall.
- Several IP telephony devices only fulfill the needs of certain markets and are often of poor quality or more expensive compared to currently available traditional telephony equipment.
- There are still problems with emergency calls ("911" calls in the United States), which must be placed locally.
- VoIP requires significant initial financial investment (hardware, software, implementation).
- Not all data networks have sufficient extra capacity/performance for VoIP.
- Existing long-term contracts for traditional PABX/PBX equipment make the migration to VoIP financially unattractive.
- As with all new systems, VoIP installations have a learning curve and will probably have initial problems.
- Potential users are confused by continuing regulatory issues in several countries of the Western world.
- There are serious concerns within the potential user base about the reliability of VoIP services. There are also concerns about the

interoperability between new VoIP installations and existing PABX systems and the PSTN.

To sum this up, one can say that, from the point of view of a significant part of the potential user base, the market is not quite ready. The quality itself is also still an issue. This is especially true for the voice quality but also for the quality of service in general. These concerns will be addressed in the next section.

4.2.5 VoIP and Quality of Service

The following steps are necessary for the transfer of voice data using VoIP (beginning and ending a connection are discussed separately):

1. Recording and digitization of the voice signal (permanently during the connection);
2. Data compression and packet building of the data stream;
3. Transport over the IP network;
4. Acceptance of incoming packets through the receiver (permanently during the connection);
5. Assembling of data packets;
6. Data decompression;
7. Reverse conversion (from digital to analog signal).

Even from this simple, nontechnical description of VoIP transmission, one can understand the possible problems that could occur during this process. To make VoIP function at the level of quality expected by the user, perfect compression and decompression of data are necessary, as well as problem-free transfer via the network. With these deliberations, we are already in the middle of a QoS discussion.

The main question that arises here is "What exactly is good speech quality?" The answer to this question is first of all a very personal one, depending on user experiences and expectations. Voice quality, as experienced by the user, depends mainly on the following factors: clarity, delay, and echo. These factors are mostly a matter of personal subjective experience. Underlying these are a few technical parameters that need to be addressed. The following are most important in user acceptance of the speech quality:

- Bandwidth;
- Latency;

- Jitter;
- Packet loss.

The bandwidth necessary for voice traffic is normally relatively low. The user typically experiences just 64 Kbps as premium quality. Other speech implementations, for example, the one used for GSM phone service are below this but also deliver acceptable quality from the user's point of view.

To transfer voice traffic over IP networks it must first be digitized. This analog-to-digital conversion is done in a standardized way by using different codecs (*codec* is an abbreviation constructed from CODer and DECoder). The bandwidth needed for transmission depends on the codec used. The main requirement for a codec is to keep the required bandwidth low while also ensuring good voice quality. A codec also helps with compensating for packet losses (forward error correction) and buffering packets and bringing them back in the right order if necessary at the receiving end of communications.

For the transmission of voice traffic, several alternative codecs are available. Typical bit rates range from 8 to 64 Kbps. Even the 8-Kbps codec named G.729 (A/B) provides sufficient speech quality to clearly understand every word spoken.

For voice transmissions, other factors besides data compression matter even more for connection quality. High latency time is a serious problem for any kind of bidirectional communication and therefore also for telephone conversations. Latency is the total sum of the time used for encoding the voice at the sender, the time needed for packet building of the digitized voice, the time necessary for transmission within the network and for buffering at the receiver, and finally the time necessary for decoding (see Figure 4.3).

Although call signaling in VoIP telephone installations, for example, with the H.323 standard, is done by using a so-called gatekeeper, the connection between the end devices occurs directly via the Real-Time Transport Protocol (RTP), which is based on IP.

There are two major benefits one can experience here: By connecting directly with each other, the telephony server will not experience shortages of computing resources and signal latency will be as low as possible, since voice data can take a direct path without taking the long route via the telephony server. If the total latency between sender and receiver (the *one-way latency*) is lower than about 100 to 150 ms, the user will not experience the delay as troublesome and will accept the call quality. Between 150 and 200 ms, the delay will be clearly noticeable by the user and will sometimes be seen as bad quality. With 400 ms or more, a normal two-way conversation becomes impracticable. The conversation partners would cut each other off in such an unfriendly calling environment. The ITU G.114 rule recommends a maximum of 150 ms. A

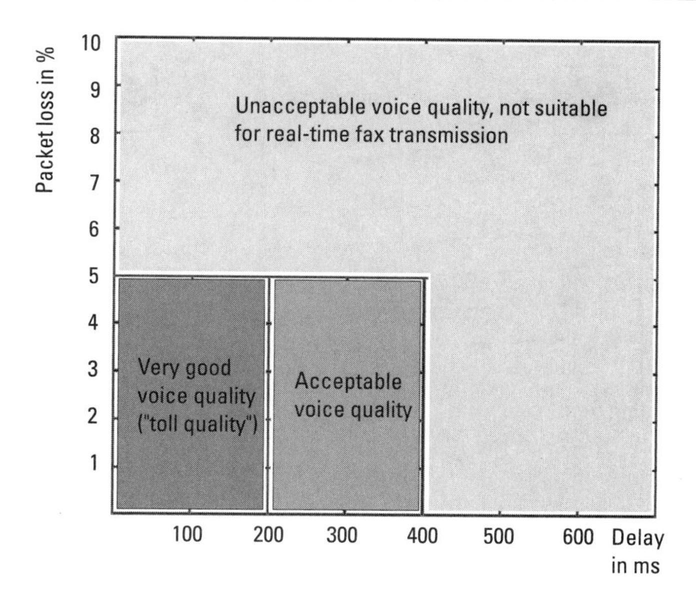

Figure 4.3 Voice quality in relation to delay and packet loss. (*After:* [4].)

simple one-way communication will not be affected by latency delays. For example, listening to an online radio station is not affected by any amount of delay, since there is no interaction. One will probably only discover that a delay exists when comparing a local clock with the station time check or the beginning of an hourly news broadcast.

Another quality factor is jitter (see Figure 4.4). Jitter is heard as a signal distortion caused by differences in the delays of incoming packets. Jitter happens because not all packets require exactly the same amount of time to travel over the network from sender to receiver. To fix this, data packets are normally buffered at the receiver. Because buffering prolongs the latency time, there is little room for this. Modern software allows dynamic buffer size allocations: When jitter is low (packets come in at regular intervals), the buffer size will be minimized. If intervals become more and more irregular, the buffer size will be increased. Intelligent software mechanisms are used to find the right balance between jitter buffering and signal delay.

Packet loss is another problem that must be addressed when transmitting voice traffic on the net. Speech reacts very sensitively to packet loss. The actual level of nuisance depends on the following factors:

- Codec used;
- Packet size;

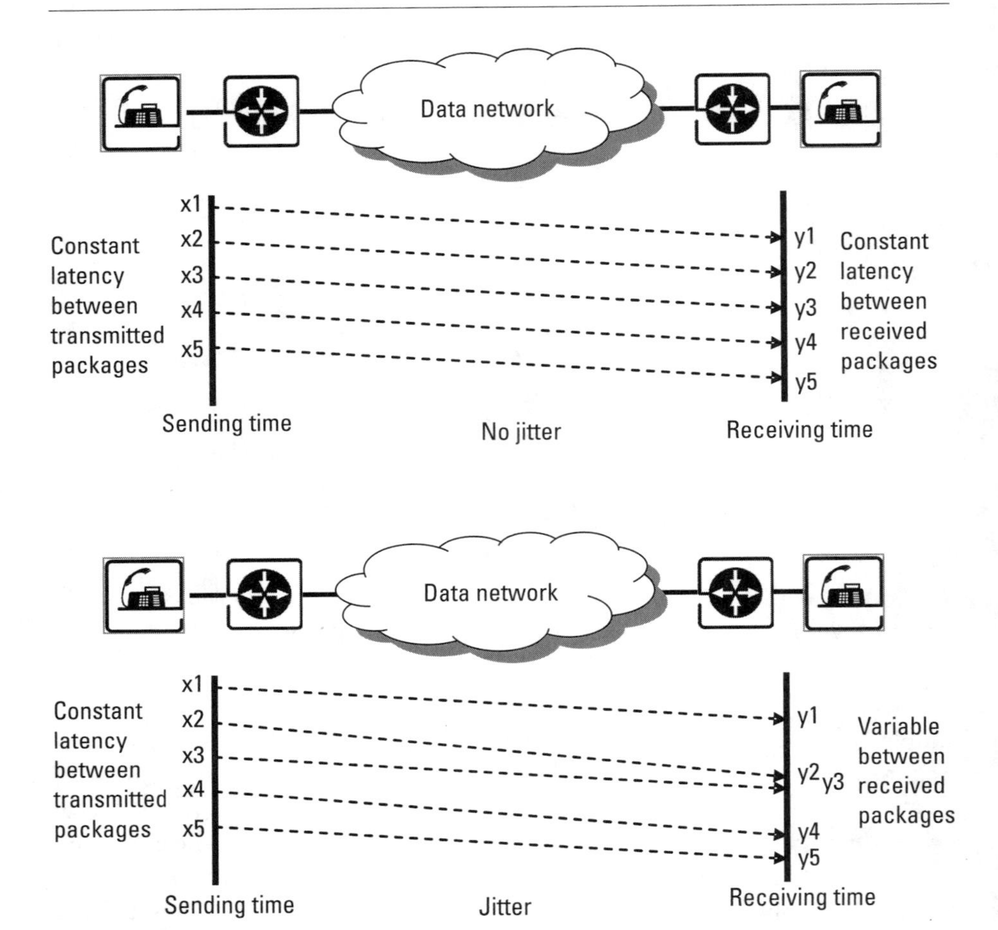

Figure 4.4 Jitter.

- Availability of error correction;
- Loss pattern.

As a rule of thumb, one can say that, depending on the influencing factors just listed, even a 1% to 5% packet loss rate can be enough to spoil the user experience.

Also, the acoustic quality of VoIP end-user devices is sometimes not as high as the user might expect since it is influenced by many factors (see Figure 4.5). Cheaper white-label IP phones in particular are often of very poor construction quality, especially when it comes to important parts relating to acoustic quality (such as speakers, microphones, and signal processing units).

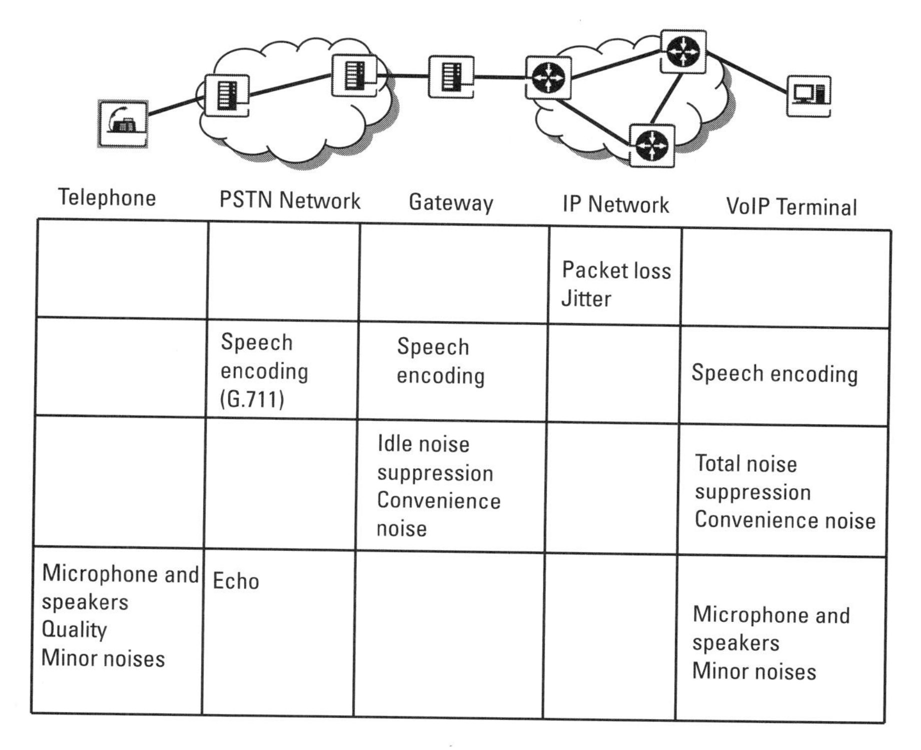

Telephone	PSTN Network	Gateway	IP Network	VoIP Terminal
			Packet loss Jitter	
	Speech encoding (G.711)	Speech encoding		Speech encoding
		Idle noise suppression Convenience noise		Total noise suppression Convenience noise
Microphone and speakers Quality Minor noises	Echo			Microphone and speakers Minor noises

Figure 4.5 Possible influence on the speech quality. (*After:* Swyx Communidcations: IP Telephony Guideline.)

Another problem that comes with any telephone is the occurrence of echoes. Echoes will only be recognized as such when the latency is above about 20 ms. ITU standard G.113 lists 30 ms as maximum. Echoes can be subdivided into two categories:

- *Line echo:* only occurs when dialing into the PSTN. Special measures can be taken against this at the transfer point between the IP network and the PSTN.
- *Echo from terminal equipment and from hands-free kits or speakerphones:* These are more difficult to fight. The best idea is to avoid this problem by using devices of higher quality.

Quality can also be compromised in other ways, especially if one or both communication endpoints are connected via the PSTN.

VoIP quality is not only defined by the audio quality of the connection; reliability is also a major issue since there are various new threats (see

Figure 4.6). As described earlier, the user of a telephone service expects a dial tone after lifting the handset from the cradle. He probably knows about call interruptions or "network busy" from previous experience with mobile phones, but he will expect any fixed-line phone to function reliably in the accustomed manner. It is not easy to describe these expectations in a monthly or yearly availability number, but 99.999% availability (which is a maximum of about 5 minutes of downtime per year) will come close to what he knows from the PSTN. This 99.999% value is an almost magic value when it comes to reliability discussions and is sometimes just called "five nines."

All discussion on percentage values aside, general redundancy can help to reduce system outages.

User acceptance still largely depends on reliability. This is not only true for company users, but also for private households. The Mercer Management Consulting firm [5] found this assumption to be true in a survey conducted in the United States and the United Kingdom in 2004. Only 6% of all private users asked were ready to migrate to VoIP for reasons of cost (lower monthly costs/low or no usage costs). The acceptance only rises when additional needs are important (reliability, voice quality, numerical portability). Some users also wanted additional features such as SMS at no extra cost in order to justify the migration.

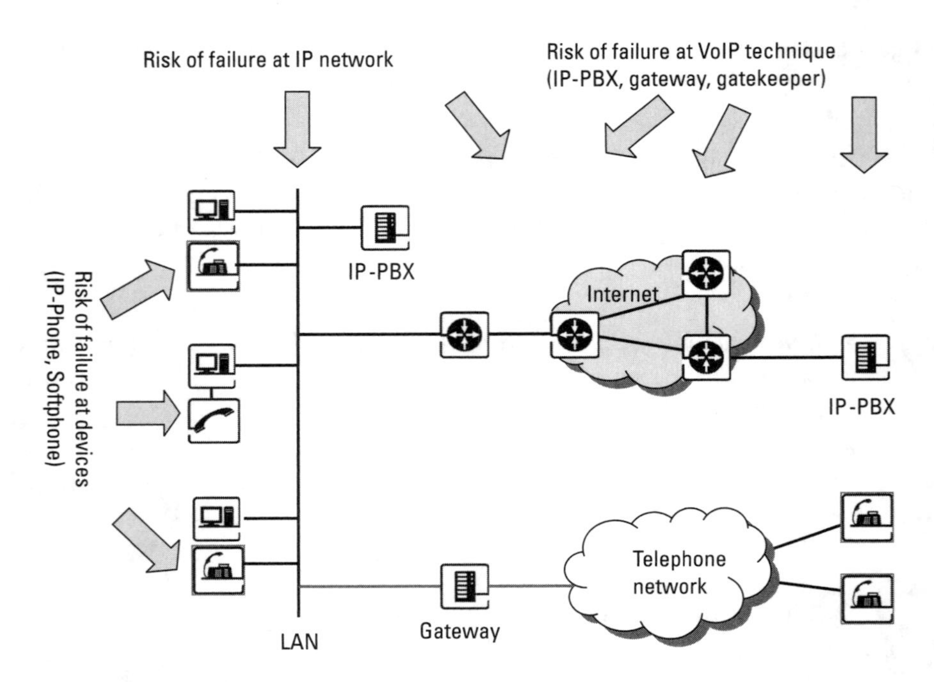

Figure 4.6 Risks/possible sources of malfunction with VoIP.

From another point of view, one can say that the lack of quality is a barrier for the acceptance of VoIP. A telephone service where quality is below the service experienced in the PSTN will not be accepted. Voice quality is the most important attribute: 37% of the users had this opinion, while 33% viewed connection breakdowns as unacceptable.

To really gain ground, VoIP must live up to its promises and must deliver the quality the user already expects from the existing phone service.

It is simply not enough to transmit voice traffic on the cheap. Intelligent mechanisms that secure the quality the user expects must be available at the same time. To deliver real "end-to-end" quality, VoIP needs additional technology. One of these is called MPLS and is described in detail later in this chapter.

4.2.6 VoIP Standards

Different standards can be used as basis for implementing VoIP: H.323 and—even more important—Session Initiation Protocol (SIP). Both standards were defined by supranational organizations. H.323 is a standard provided and defined by the International Telecommunication Union (ITU), whereas SIP was defined by a working group of the Internet Engineering Task Force (IETF). Both standards support VoIP but are constructed in completely different manners.

H.323 Standard

H.323 can be seen as a standard for all packet-oriented multimedia communications and is based on standards for video communications that have already been used for several years. H.323 describes devices (terminals), hardware, software, and services for multimedia communications via an IP network but without defining service quality.

H.323 contains the following components (called *entities* within the written standard description):

- Terminals;
- Gateways;
- Gatekeeper;
- Multipoint control units.

H.323 terminals can be integrated into a PC (as software applications) or be provided as stand-alone devices. The minimum requirement for terminal features is audio transmission capabilities; video capability is optional.

A gatekeeper can provide the following functions:

- User administration (name, numbers, rights, e-mail, and so forth);
- Authentication of users;
- Translation of phone numbers and e-mail addresses into IP addresses during connection start-up;
- Generation of log files for call data (length, connected users);
- Administration of central phonebooks;
- Bandwidth management;
- Replacement administration for clients who are switched off or not connected to the network;
- Connection management (to connect to several gateways);
- Web front end for system management.

The multipoint control units (MCUs) act as central systems for managing multipoint connections.

SIP Standard

A simple but powerful alternative standard is the Session Initiation Protocol. The SIP standard is defined—like all standards established by the IETF—as a request for comment (RFC), here RFC 3261 (originated in June 2003; see http://www.ietf.org/rfc/rfc3261.txt for the text of the original standard). The SIP standardization activities that began in 1999 had the goal of establishing a simpler, more modular protocol for VoIP than the complex H.323 standard.

SIP is a text-based protocol, similar to the well-known Hyper Text Transfer Protocol (HTTP) or Simple Mail Transfer Protocol (SMTP), for the initiation, modification, and termination of interactive communication sessions between two or more participants. These sessions include multimedia conferences, Internet phone calls, and multimedia distribution. The transfer itself relies on RTP (see Figure 4.7).

A SIP session consists of the following phases:

- User location;
- User availability check;
- User capability check;
- Session setup;
- Session management, including transfer and termination.

SIP is a very flexible protocol, can be used with different transport protocols, and does not even require a secure connection. A SIP client can therefore also be implemented using the simple User Datagram Protocol (UDP). SIP

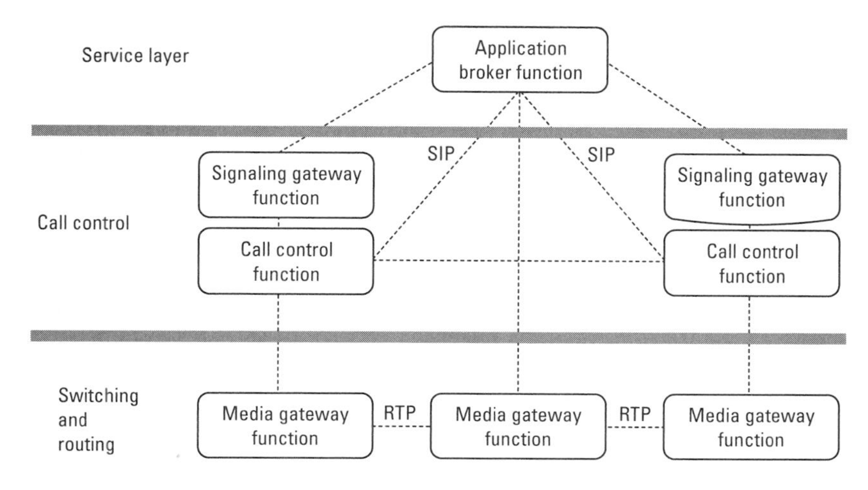

Figure 4.7 SIP overview.

includes a Web-based address book and supports call transfer as well as dynamic localization.

The H.323 and SIP standards are compared in Table 4.1.

Table 4.1
Comparison of VoIP Standards

Attribute	H.323	SIP
Published by	ITU	IETF
Extent of content	1,400 pages	250 pages
Spread	Widely spread	Upcoming
Compatibility with PSTN	Yes	Partial
Architectural model	Monolithic, closed	Modular, open
Completeness	Complete protocol stack	SIP only covers connection setup
Call signaling	Q.931 via TCP	SIP via TCP or UDP
Message format	Binary	ASCII
Media transport	RTP/RTCP	RTP/RTCP
Conference calls	Yes	Yes
Multimedia conferencing	Yes	No
Addressing	Host or telephone number	URL
Call termination	Explicit or TCP release	Explicit or time-out
Encryption	Yes	Yes

A decision to implement VoIP is, of course, a decision for implementation based on a standard. Interoperability between VoIP implementations based on different standards is possible via gateway applications.

4.2.7 Security Aspects of VoIP

Voice networks based on IP are, as discussed earlier, the future of voice traffic but, when it comes to security measures, they require a different approach than traditional voice networks and implementations to prevent security breaches from occurring.

Several new potential security risks exist that were unknown in traditional voice networks and include denial-of-service attacks, worm threats, and voice spam.

As with any other IP-based service, security problems are common. From a pessimistic viewpoint, one can expect that most security problems already known in IP networks will make their way into the VoIP world, and VoIP maintenance will probably consist of fixing problems, installing patches, and fighting attackers. Service providers as well as software vendors emphasize that security and reliability of VoIP services and software will be as good as that of traditional telephony implementations.

Apart from the expectations derived from experience already gathered elsewhere, one can examine the concrete technology more closely. One will find that eavesdropping on connections, especially when it comes to WLAN telephony, will probably be a problem because existing protocols give little, if any, protection against any unauthorized eavesdropping and recording. Strong encryption is currently not supported by most stand-alone devices due to the lack of computing power but of course will not be a problem when it comes to soft phones (phone terminal implemented using software on a PC). The upcoming Secure Real-Time Protocol (SRTP) standard is designed to solve this problem.

Another key problem that will probably need to be addressed is identity theft. This includes identity theft with the intention of stealing services (such as unauthorized usage of VoIP–PSTN connections) as well as with the intention of manipulating the identities of callers, with consequences for service availability and confidentiality for the end user.

The consequences for the company using the service, or having its own VoIP implementation, as well for the service provider can be service outages (with resulting financial losses) as well as financial losses due to unauthorized access to paid services (such as PSTN gateways as well as 1-900-type service phone numbers).

Although nobody likes to talk about it, service theft is not solely a problem with VoIP alone; it has been and still is a problem in traditional telephony systems.

When it comes to VoIP security discussions, open-source VoIP implementations remain a white spot. Although developers confirm the security of their implementations, this must still be proven in the real world. Given the current problems experienced with LINUX-based software, a number of question marks remain. Comparisons with traditional phone services are difficult since there was no such thing as a free PABX/PBX system in the old world.

Permanent logging of any connection and extraction of statistical data from the log files can help with early detection of fraud. This activity is sometimes provided as a service (fraud management). Insurance against financial losses caused by telephony fraud is even offered by some vendors. Each user must decide for himself or herself if this insurance is necessary but there are measures that every user should take. Every relevant system-specific security feature should be used. These can include segment-oriented device filtering, MAC address surveillance, and the deactivation of any "auto connect" features provided by the software vendor or service provider. Auto connect allows easy integration of additional phones but also makes unauthorized connections to the service easier.

4.2.8 Implementation of VoIP

When it comes to the implementation of VoIP, a lot of technical decisions must be made. VoIP implementations have different characteristics that depend on whether the provider or vendor has a background as a PABX developer/producer, a service provider, or as a network component vendor.

One should closely examine certain aspects before beginning construction or purchasing of your own VoIP Infrastructure. First of all, the network must be suitable for supporting VoIP. Based on the small bandwidth needs for every single phone call, this is normally no problem in standard 10/100/1,000 Mbps networks. Difficulties arise when it comes to securing the quality of service. For voice traffic this means prioritization with respect to other network traffic, so that large file transfers or other network action will not hinder voice communications or compromise voice quality. Because normal IP networks do not include prioritization mechanisms, new technologies such as MPLS are needed to solve this dilemma, especially in the corporate WAN. Data traffic in MPLS is classified into different classes of service (CoS). These different classes are treated differently. For example, voice and video traffic has a high priority, and file transfers have low quality and therefore sometimes have to wait or will be slowed down if bandwidth is insufficient to simultaneously support all requested services. Since nobody wants to compromise the performance of key network applications such as ERP, one must be careful when fine-tuning the CoS specifications to best suit all of the user requirements. This requires time and expertise.

Upgrades to the infrastructure sometimes go hand in hand with VoIP implementations. Most plans include estimations of the amount of *on-net traffic*, which is relevant for bandwidth needs. On-net traffic includes all calls that are transmitted through the company's network infrastructure, whether it is LAN or WAN while off-net traffic includes all calls (or parts thereof) that travel over someone else's network infrastructure.

A procedure called *application profiling* helps with determining voice traffic but also with any other traffic driven by applications on the net. Application profiling includes a directory of all applications that are already in use or plan to be used within the planning horizon. It includes a usage profile with average data traffic, peak data traffic, and other data traffic characteristics for each of the applications (see Figure 4.8). For VoIP one would calculate the expected call quantities and durations as well as the number of parallel calls during peak and

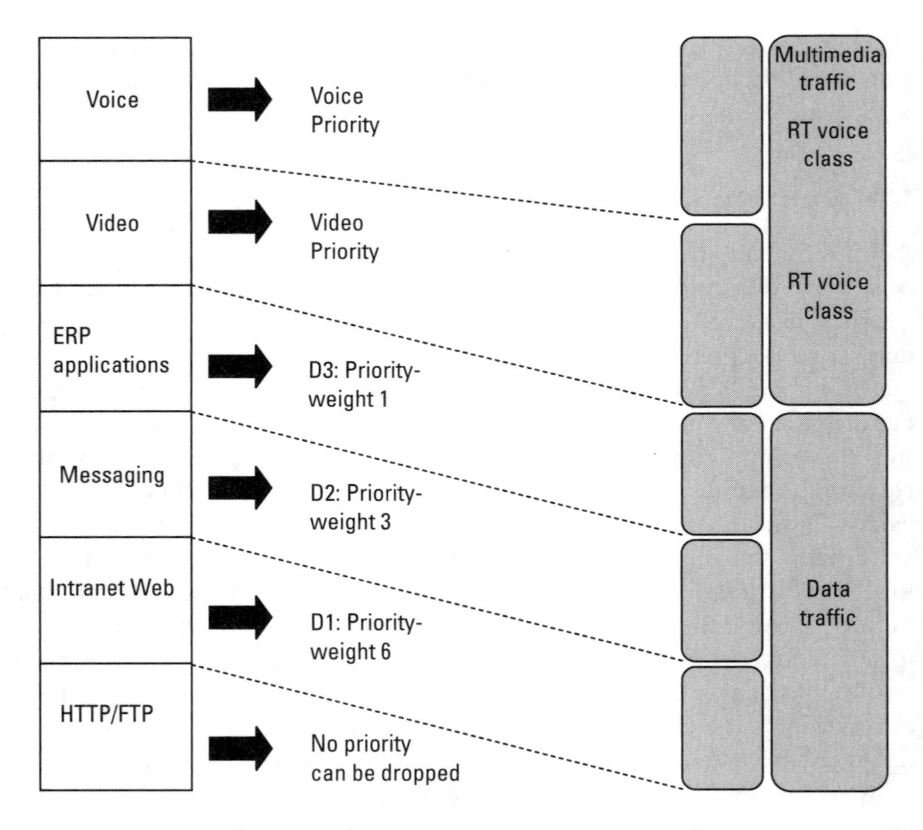

Figure 4.8 Quality-of-service classes (according to the Yankee Group).

off times but also any other related traffic, such as that expected from other VoIP applications such as voicemail. These numbers are integrated within a *bandwidth budget,* as the calculation is sometimes called.

Within these calculations one must pay special attention to the corporate WAN since the available bandwidth usually is much smaller than in a LAN. Well-known methods for planning PABX/PBX systems and PSTN connections help us obtain the necessary data here. These methods include not only voice volumes (number and duration of calls), but also the distribution of incoming and outgoing calls within the time of day—sometimes down to the minute.

Of course, not all network planners go into this level of detail. Very often one only uses rules of thumb derived from traditional knowledge, or simplified calculation schemes based on the number of phone devices and the typical number and duration of calls over 1 hour. As already mentioned, the number of on-net calls must be weighted against the number of off-net calls to obtain results that are useful as a budgeting basis. Other aspects such as compression ratios or the number of calls ending with "no one available" are very often not included in the calculations.

A VoIP calculation model [6] presented by the analysis company MetaGroup in 2004 came to the following conclusions (under the preconditions that there is a 30% share of on-net traffic of the entire amount of traffic and every phone is used for 6% of the total time):

- 25 phone devices lead to a bandwidth requirement of 50 Kbps;
- 100 phone devices lead to a bandwidth requirement of 151 Kbps;
- 200 phone devices lead to a bandwidth requirement of 274 Kbps.

If the share of on-net traffic increases to 40%, the bandwidth needed for 100 telephone devices would increase to 202 Kbps. These numbers give a first impression of what to expect when it comes to voice traffic over the network.

Of course, there are several other aspects besides the bandwidth. Fragmentation of packets is one of these. Because voice packets are small, there are usually no problems to be expected when regarding voice alone. But when other traffic comes into play, the packets used are relatively large. If total bandwidth is already relatively low, such as upstream from DSL connections (often used for connecting home offices), this can lead to conflicts and compromise voice quality. Because most routers support fragmentation of packets down to smaller sizes, this problem is solvable.

Another aspect is the way call processing works. There are two major options here: centralization versus distribution of the necessary infrastructure. This question also touches other aspects (besides the core functionality of allowing phone calls to work like voicemail). If one must decide between centralized and distributed call processing, one should contemplate the consequences that a loss of WAN connection between the branch office and the central office will have: If centralized call processing is used, the phone service will be down at the branch office. There even is a special term for this aspect. It is called *remote survivability*. A solution for the challenge of keeping phone services up throughout the corporate infrastructure when losing a connection between the headquarters and one or more branch offices would be distributed call processing or backup connections for the WAN, which for example, are supported by some router types as a standard feature.

Besides call processing, the power supply can also be a problem with VoIP installations. This is no problem with standard PABX/PBX systems. A single emergency power supply for the PABX/PBX will also power the phones connected to it, while with VoIP every phone will usually have a separate power supply and therefore will also need a separate emergency power supply to keep it working during power outages. Of course, nobody would do this and we will not discuss workplace aesthetics here. This also will reduce the cost balance of VoIP and could kill all cost benefits derived from the decision to use VoIP. VoIP phones that are powered over the Ethernet connection could easily solve this problem but special routers are required for this and these would also need emergency power supplies (as with every VoIP installation, but unlike a traditional PABX/PBX phone service).

Finally, the connection with the PSTN is an important aspect to consider. Which connections can be cut off, which connections will still be needed, and which connections will be needed as future additions to make the best out of VoIP?

Off-net traffic (as mentioned, connections to devices that are not part of the company's own infrastructure) must be considered here from a cost and capacity oriented point of view. Besides this, the connection and call routing to emergency services (911 calls) are a problem that still remains to be solved: If a company has its headquarters in town A and branch offices in towns B and C, and an infrastructure based on centralized call processing with one centralized PSTN connection at the headquarters, then any emergency call originating from within the corporate phone system will reach 911 in town A, although help is actually needed from somewhere else (town B or C in this case). There is still no simple solution to this system-dependent problem of VoIP. Regulatory issues relating to this topic are still being discussed worldwide. A simple warning sticker at the phone device that says "911 cannot be called here" was one of the last suggestions in 2005 to help in solving this problem, along with vendor attempts to escape from regulatory spheres of influence by claiming that their products are not phone services (which would clearly lie under regulation) but data services (which in some countries are not regulated).

To provide stable VoIP service, several aspects described in the following list should be examined:

- *Determination of requirements.* For each company location, a detailed documentation of the telephone infrastructure should be made, including not only phones and numbering plans but also all functionality such as call forwarding and voicemail. If it is technically feasible, not only should the features that are available be documented, but also the usage intensity. The knowledge gained by doing so delivers the base requirements for the new system and its necessary features.
- *VoIP readiness test.* At each location the cabling and network components should be checked to see if they are suitable for VoIP usage. Available space for any additional equipment, as well as the availability of power outlets (for non-Ethernet powered phones) should also be checked. Performance over the WAN should also be included in the VoIP readiness test.
- *Standardization.* Although there are standards for VoIP, as discussed earlier, not all devices and components are interchangeable yet. Vendors are still trying to sell their own proprietary equipment with limited compatibility.
- *Development of a new centralized numbering plan.* Existing numbers should be integrated as far as possible, to avoid user and customer confusion, and to limit the required changes (for example, on business cards) only to the cases where it is unavoidable.
- *Organizational integration.* With the implementation of VoIP, telephony moves at least partly into the IT department. (If this has not already occurred, some corporations still have a separate department for telecommunications or see the procurement department as being responsible for telecommunications.) Service and administration processes need to be changed and if necessary adapted to the new management structure within the IT department to make sure that responsibilities are met with the new structure.
- *Pilot project.* A pilot project at one or more company locations with a minimum duration of 1 month or more is often recommended. Within this pilot project, performance evaluations should be done to make sure the initial assumptions and calculations are correct.

As mentioned earlier, the user has high expectations regarding the availability and quality of corporate phone services due to past experience. Providing the expected quality is the major task for any successful implementation of VoIP. The consequence for monitoring and performance evaluation is that information about lost or broken connections is more important than typical performance parameters as used in traditional tests.

4.2.9 Use of VoIP

A provider network-based VoIP implementation can be an alternative option to a company-based installation. Similar to CENTREX, which provides PABX/PBX-like services using the provider network infrastructure, several providers offer similar services using VoIP. These services are sometimes marketed as *IP-CENTREX*. The advantages of IP-CENTREX services are similar to those of the traditional CENTREX systems:

- No costs for purchasing, administration, and servicing of PABX/PBX systems (depending on the tariff models offered);
- High scalability;
- Good flexibility with regard to additions and changes;
- Simple realization of telephone implementations integrating more than one company location;
- Easy integration of teleworkers and even mobile workers.

Almost every service of this kind comes with a Web-based application to allow changes by the user or the user administrator. Some vendors have differentiated these functionalities to suit the needs of up to seven different groups having different usage rights. These applications then act as components of provider-offered service portals or offer portal functionality on their own.

Besides IP CENTREX, the term *hosted voice* is seen in provider offerings. Although these two terms are often used synonymously, there is a distinction. IP-CENTREX delivers standardized, off-the-shelf services that also cater to small and medium-sized companies, whereas hosted voice describes more individual services catering to the needs of larger corporations. The technological basis remains the same: No equipment (except the phones) at the client site belongs to the client, and the provider manages all software/hardware components within its own infrastructure. The analysis company Gartner describes these services as "in-thecloud" services, in reference to the provider network. In-the-cloud services are not limited to VoIP implementations, but can also be integrated with other services such as managed security.

4.2.10 Advantages of Integrating Voice and Data Networks

Until now, the adoption of convergence technologies within corporate networks was slow. The reasons lie very often with the lack of knowledge of the positive effects that come with implementation but also in misconceptions of the quality, reliability, and functionality of the systems and services available on the market. Fast adoption is also hindered by the divisions between the company departments for voice networks and data networks, which still exist in some companies. However, the benefits are undeniably there and can be sorted into cost savings as well as flexibility increases. The latter can also lead to increases in productivity.

Measurable Cost Savings

The main driving influence for voice-data integration projects is a measurable saving on the total costs. Although TCO and ROI calculations provided by software vendors, service providers, and consulting companies are often seen for what they really are—advertising—nobody can deny the cost-savings potential that lies therein. As early as late 2002 [7], one study calculated usage cost savings of up to 49% and total cost savings of up to 32% compared with traditional telephony in companies with more than 500 employees. Similar results can be seen in studies from other researchers. Where do these savings come from?

- *Lower infrastructure costs.* A unified network means that only one cabling structure is necessary for voice and data. Because cabling is typically already provided for voice as well as for data, significant savings are only achievable with new implementations in new offices or at new locations, so-called "greenfield" installations. As mentioned earlier, power supplies must be provided for VoIP, whereas standard telephones connected directly to the provider and the PABX/PBX do not normally need a separate power supply. The best solution: The infrastructure must support "power over Ethernet." The alternative would be a separate power supply for every phone, and it is not easy to explain to the user that this is required by the "new" telephone system.
- *Bandwidth savings.* The move from two networks to only one automatically leads to lower bandwidth requirements since voice and data traffic have different patterns and different usage scenarios. In total, a lower bandwidth than previously required can be expected for a convergent network. Voice compression also reduces the amount of traffic and leads to lower related costs.
- *Lower usage costs (lower costs per call).* Calling costs, especially when it comes to long-distance calls and international calls, are another core

factor in cost comparisons between traditional and convergent network structures. Companies having locations throughout the country and in other countries reap the most benefits from this, since they can route the calls—as long this is useful—within their own networks. Within this infrastructure, the PSTN gateway located nearest to the destination of the call will be used (within the same town, area code, or country). Instead of long-distance and international calls, cheap (or even free) local calls are very often used. These savings can also be realized in part without having your own WAN network. Most providers offer this kind of cost-saving call routing service within their own network instead of using the corporate WAN.

- *Lower administration costs.* It is easy to understand that the expected administration costs for one network are lower than for two networks. Because the structure is simpler, there is less administrative work and therefore lower costs.
- *Reduced relocation costs.* The most popular topic in any discussion about the cost benefits of converged networks/VoIP is the reduction in relocation costs. Relocation of user devices is indeed simplified as well as the removal and addition of devices. The sum of the savings achieved depends on the development dynamics of the company. An expansion-oriented company will profit more than a traditional company with a very static structure (and therefore less need for relocation).

Apart from the topics already listed, there are other benefits that are more difficult to measure in terms of direct cost savings:

- *Productivity increases.* The convergence of telephone, fax, and e-mail applications makes it easier for users to communicate and speeds up access to relevant content and messages.
- *Improved functionality.* VoIP delivers added functions, in which users can benefit from various usage scenarios, such as e-commerce systems and call centers.
- *Vendor independence.* Due to the usage of vendor-independent standards, the level of interoperability increases and the telephony infrastructure can be composed from a broad product portfolio (at least if the vendor implements the standards correctly).

Increased Flexibility

More flexibility is a central company requirement (as addressed in Chapter 3). The ability to rapidly adapt to changing needs is worth something, although this is hardly measurable. The consequences are reduced costs and time frames necessary for relocation as well as new applications:

- *Unified messaging.* Unified messaging allows access to all communications (telephone, fax, e-mail, video messages) in one single access and allows a more structured approach to working.
- *Multimedia collaboration/videoconferencing.* Multimedia collaboration/videoconferencing allows cooperative work between teams that are distributed across several locations and cooperation/dialogue with suppliers and buyers. With convergent networks, one expects a breakthrough here, which was expected by now but has not happened due to the high costs and difficult provisioning of such workgroup-supporting applications and services.
- *Teleworking.* By making use of low-cost broadband connections for private households (based on cable modems and DSL), integrated communications can also be used within the teleworkers' home offices. This includes unified messaging as well as call center services.

Better Customer Service

Converging communication channels on a unified platform can also help to deliver better customer service. Benefits begin when data for customer interaction must be collected and collated. Benefits also arise during usage, for example, from common usage of a Web site where a customer reads a Web site or goes through an ordering process with a call center agent while simultaneously having a phone dialogue with them.

The call center itself can easily be extended to cope with teleworkers as well as with employees at other branch offices. This all can happen on the fly as soon as it is required, for example, in response to a short but intensive advertising campaign. Within the call center, localization of calls with forwarding to a call center agent responsible for this special area is also easier. But customer self-service can also be improved, because Web and voice services can be offered on the same system platforms.

4.3 Rich Media and Unified Communications

The current discussion on VoIP sometimes makes us almost forget that there are other applications besides voice, and that there is already a slow but steady development toward what is called *rich media communication.* This term includes streaming audio and video but also Web conferencing/instant messaging for corporate usage (training, communications) and usage within customer communications (marketing, advertising, service and support).

The most important uses of streaming audio and video include employee training and broadcasting of corporate events (such as shareholder meetings). In large companies streaming video already is the dominant technology for so-called "business TV," corporate television broadcasts where the viewers are employees.

Rich media includes audio, video, and data. Typical for rich media communications is the realization of multipoint connections by transmitting data to several users at the same time. Rich media communications are often seen as an extension of VoIP. Since, in part, the transmission protocols are all the same, this view is easy to understand. A separate opinion on this topic is also useful, since the bandwidth required is much higher with video transmission than with voice.

If one wants to transfer a videoconference supplied in the typical resolution of today's desktop or notebook screen (XGA = 1,024 × 768 pixels), the data rate required is 283 Mbps (see Table 4.2). Without data compression, video transmissions can hardly be done using current networks, even when the resolution is reduced, for example, to VGA (VGA = 640 × 480 pixels). While the audio compression rate is typically around 10:1 or better, the video compression rates are typically much higher (about >100:1).

The Moving Picture Experts Group (MPEG) defines several standards for the compression of moving pictures that also include variants for low bandwidth communications (see Table 4.3).

Table 4.2
Image Formats and Required Data Rates (at 15 fps)*

Format	Width (pixels)	Height (pixels)	Data Rate Without Compression at 15 fps (Mbps)
QCIF	176	144	9
CIF	352	288	36
VGA	640	480	110
SVGA	800	600	173
XGA	1,024	768	283
SXGA	1,280	1,024	472

* fps = frames per second.

Table 4.3
Comparison of Video Compression Algorithms

Format	Quality
MPEG-1	Up to 2 Mbps, limited picture quality
MPEG-2	Up to 80 Mbps, up to 60 fps, higher resolution
MPEG-4/H.263	Model-based compression with quality losses, suitable for low bandwidth, 4.8 to 64 Kbps
MPEG-4 AVC/H.264	Increased quality with lower bit rates (PSNR 50% better than H.263, but needs high computing power for real-time compression; for standard TV resolution, a processor with more than 2 GHz is necessary)

IP conferencing and rich media applications are expected to grow during the next few years. Significant drawbacks, however, will hinder usage: Problems exist with firewalls as well as with network address translation (NAT), a technology for translating between public and private IP addresses at the border of the public network and the private network.

There are high expectations for visual data networking. Increasing bandwidth within the corporate networks as well as codecs with better audio video compression support this development.

As with voice alone, the implementation of rich media in the corporate network requires a special focus on quality. While rich media focuses largely on applications, the concept of unified synchronized communications (USC) lays its focus on integration and collaboration between different communication needs thorough integrated administration and intelligent usage:

- Availability management (for incoming communications);
- Finding of users and intelligent call forwarding;
- Starting and usage of multimedia conferencing features via other media, for example, during a phone call.

USC will also integrate existing collaboration concepts. The goal of USC is to increase employee productivity, to reduce delays in communication, and to increase service flexibility through simplified usability. Several similar concepts that share all or some of the concepts listed above are also currently in discussion.

4.4 Multiprotocol Label Switching

4.4.1 IP Is Not Enough

Despite the rapid development of IP as the de facto standard for data networking, there are limitations that question the use for all applications requiring special service quality. The previous chapter already described the importance of a defined quality for voice traffic over the net.

Routing within IP networks works in a "connectionless" manner, so a defined quality of service cannot be made available. Every router within the network decides independently about the forwarding of every data packet by analyzing the destination IP address and the knowledge about the network environment. Packets are received via the incoming port, are administrated by the "forwarding engine," and leave the router via outgoing ports.

This process is based on the so-called "forwarding information base" (FIB) within the router. The decision over packet forwarding is derived from the content of this database, which contains information about the network neighborhood.

Each packet that comes to the router is examined based on the target address. The database of the FIB is searched and then, according to the results of the search, the packet is then forwarded on the correct outgoing port of the router to the appropriate network segment. Although this appears complicated, this was and still is the most important way to operate, even with a strong increase in Internet traffic. In summary, one can say that this kind of on FIB-based packet forwarding is of course much slower than any switched connection. Several years ago there were worries among vendors about not being able to cope with the rise in data traffic due to limitations of the amount of data the router can cope with in a given amount of time. So, back in 1994, Toshiba presented the idea of a methodology called *cell switching* to the IETF. Other network hardware developers presented similar considerations thereafter. Every one of these suggestions included supplementing the packet contents with an extension (a *tag*). A description of the route through the network should be integrated within this tag, based on the ideas presented. Every router on the way should forward the packet based to the information supplied in the tag and not by asking the FIB.

The concept of MPLS is based on these ideas. Although the original concept was aimed at increasing data transfer rates within routers, it can also be used for offering services with a defined QoS and therefore gives providers the opportunity to offer additional services.

Over the years one has had to live with the disadvantages of IP. Because the number and importance of applications based on IP is growing within companies and because most business critical applications rely heavily on it, the demand on bandwidth and availability is growing strongly. Standard IP technology finds its limitations here, since the concept of "we do not care where the packets go as long as they find their way" does not work well under the circumstances of present business environments.

4.4.2 ATM as a Dead-End Road?

Since the 1980s, ATM technology has been propagated as a service with defined quality and was widely used worldwide in the WANs of large corporations. ATM works on a connection-oriented basis and supports QoS with different traffic types (data, voice, and video). But ATM has severe disadvantages: High complexity and expensive provisioning has led to limited acceptance in the total market share.

Just a few years ago, most industry experts were sure that ATM would be a sure winner. It was even expected by some industry analysts that IP would adapt itself to suit ATM. Apart from the cost and complexity, unsolved issues with scalability have led to a different development.

4.4.3 The Best of Both Worlds?

With MPLS there is now a kind of hybrid product or "in-between" layer between the standard packet switching known from IP and the basic connection mechanism of ATM. MPLS is a concept for labeling data packets during data transmission. The direct path of the packet through the whole network is decided on with respect to the label at the entry point to the network and no longer by every individual router on the way to the destination. So every packet is transmitted on a direct path through the network (without unnecessary detours).

The consequence: Data packets move through the network without delay and reach their destination in the correct order. So MPLS allows "quality of service" instead of the standard IP method, which is also known as "best effort":

- Defined bandwidth;
- Short transmission times;
- High availability;
- Prioritization of applications.

To summarize, one can say that MPLS is a connection-oriented way of communication on top of a network that basically works in a connectionless manner.

With MPLS, most activity takes place at the edge of the network (see Figure 4.9). Routing only occurs within the routers at the network borders, the

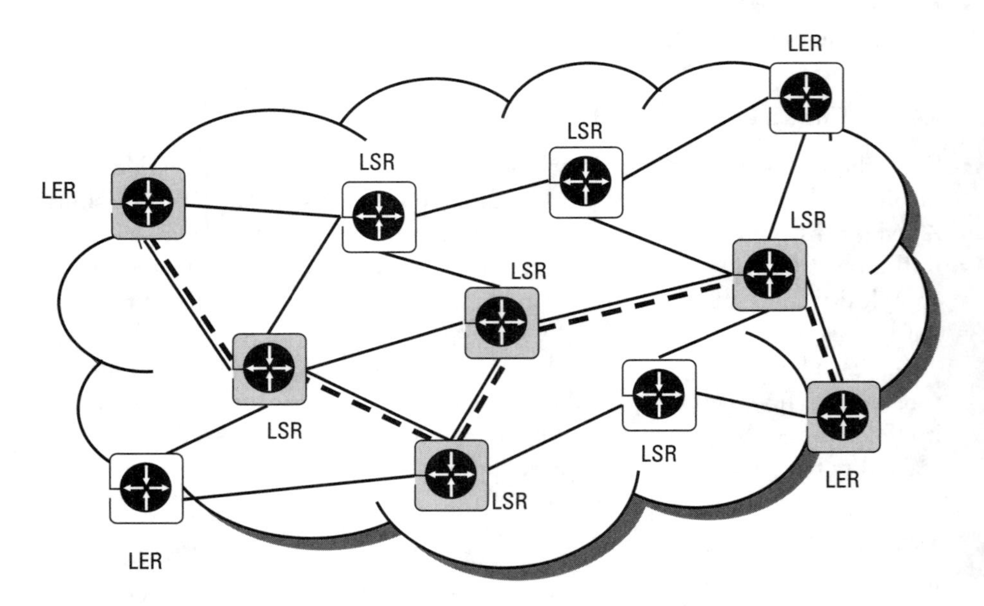

Figure 4.9 A model overview of MPLS.

label edge routers (LERs). Within the network core, forwarding is performed by the label-switched routers (LSRs), based on the labels. In general, one speaks of label-switched paths (LSPs). These paths are defined before data transmission begins. One such path gives a picture of a forwarding equivalence class (FEC). One FEC includes a group of packets sharing the same transportation requirements and which the router therefore treats equally. The decision as to which packet belongs to which FEC is decided only once with MPLS (at the time when the packet reaches the network entry point). Conventional routing makes this decision with every packet at every network hop.

LERs are routers at the edge of the MPLS network. They connect the MPLS network with other networks and play the most important role in label management. Also necessary for MPLS are the LSRs. LSRs are high-speed routers within the core of the MPLS network. ATM switches can normally be used as LSRs without making changes to the hardware.

In summary, the path that the packet takes throughout the network is fixed and will not be affected by any routing decisions of individual routers during the packet's journey through the network. With MPLS, the route taken by a packet becomes predictable and therefore QoS becomes possible. MPLS is also faster than traditional IP routing. MPLS supports ATM, frame relay, and Ethernet and has interfaces to established routing protocols, such as RSVO and OSPF.

The main disadvantage of MPLS is that to implement MPLS every router must be MPLS enabled. It is not surprising then, that network equipment vendors expect additional business here and are therefore supporting MPLS with massive marketing efforts.

For the distribution of labels, which is necessary to make MPLS work, there is no single method. There is indeed a special signaling protocol for MPLS, known as the Label Distribution Protocol (LDP), but this is not the only possible method of doing so. For BGP as well as RSVP, there are extensions designed to make label exchange work. The duties of the TCP-based LDP are described within one sentence: Build the path through the MPLS network (the label-switched part) and reserve the resources necessary.

4.4.4 How MPLS Works

The typical "how it works" list for MPLS would be as follows:

- Generation and distribution of the label;
- LIB generation in every router;
- Generation of the LSP;
- Insertion of the label, LIB readout;
- Packet forwarding.

4.4.5 Benefits of MPLS

From a technical, but also even from a philosophical point of view, MPLS is surely not the most elegant solution to providing QoS. MPLS tries to combine elements of packet switched networks with elements of line switched networks, and this leads to increased complexity within the network. But it basically works in the way it was planned and the benefits that come with it are undeniable:

- Fast packet forwarding;
- Protocol independence;
- VPN functionality (MPLS tunneling);
- Traffic engineering/QoS;
- Allowance of any-to-any connections.

MPLS and Traffic Engineering

Traffic engineering is a well-known concept to providers of line switched networks. It allows control of the traffic flow within the network to prevent potential and actual shortages, making better use of capacity and increasing the stability of the entire network. Traffic engineering within IP networks had a relative short history before the advent of MPLS. In general, the main goals of traffic engineering are as follows:

- Efficient traffic allocation based on the available resources;
- Controlled use of resources;
- Fast reaction to changes in network topology.

MPLS traffic engineering includes the following:

- Explicit routing;
- Path generation with a special view of the requirements;
- Reservation of network resources;
- Use of the protocols (CR-LDP, RSVP).

4.4.6 MPLS and Quality of Service

Resource allocation is possible due to the existence of a defined path. Bandwidth and delay become (within the boundaries of the technical conditions) controllable. For this purpose, the label includes a so-called CoS segment. This allows differentiation between requirement classes and thus allows the prioritization of voice and special applications compared to other applications or, for example, Web traffic and e-mail.

Service providers offer different classification concepts. Some companies define up to 12 different priorities, while other offerings limit class numbers to just three. Differentiation into a half dozen or more classes is certainly too much and can be seen as a marketing-driven offering, since a distinction in classes is primarily only important for voice and video (which must have first priority), then a few applications such as ERP that have second priority, and then the rest (which can be, but in most cases need not be, further differentiated).

Class definitions are the preconditions for the implementation of MPLS-based VPNs. The transfer quality can be secured "end to end" with MPLS so that applications having special QoS requirements, such as packet-based voice traffic, can be provided at the level of quality expected. But the use of service classes is not useful in every scenario; particularly when there is not yet any voice or video traffic on the VPN, prioritization will be negligible.

4.4.7 Disadvantages of MPLS

The main disadvantage for MPLS is its limited interoperability: Because no consensus has been reached among carriers about the service class differentiations, every data packet that is transferred from one provider-based MPLS network to another must be stripped down (label removed), delivered to the other network, and then again relabeled.

In an ideal world, one single carrier should be able to provide an MPLS network that is sufficient for all corporate use. This makes a large (in most cases, multinational) network structure necessary, with a sufficient number of POPs (points of presence) located near the branch office sites that must be connected.

4.5 Mobile Data Services

Mobile data services are another set of driving influences in the development of voice and data networking. Depending on the network type, connection is relatively slow but available almost anywhere (based on mobile phone installations), or very fast but only available within small areas known as *hotspots.*

4.5.1 Adoption and Extension of Traditional Mobile Phone Services

Traditional mobile phone services have been used for data transfer for several years now. The break comes with the move from merely using a mobile phone service as a line substitution to a packet switching environment, for example, as supported by the GSM standard using General Packet Radio Service (GPRS).

GSM is the most important worldwide standard for mobile phones, and it supports voice as well as data services. Basic GSM data services support only 9.6-Kbps line-oriented data traffic but with GPRS, which is an extension to GSM for providing faster, packet-oriented data transmission, data rates of up to 170 Kbps are possible (with eight simultaneous downstream channels in an area with good phone reception). Because most phones and GSM data modems only support four channels downstream and one channel upstream, GPRS is normally much slower. The response times are also relatively slow. The transfer speed can be compared with the standard modem speed for an analog phone line. Because GPRS allows "always-on" connections and is usually priced by data transfer volume, and not by time used, these benefits make it a good partner for corporate usage, for example, for an always-on connection between a mobile e-mail device and a server. RIM's Blackberry service relies largely on GPRS.

4.5.2 3G Data

Third generation mobile phone systems such as Universal Mobile Telecommunication System (UMTS) not only include features such as video telephony but also bring broadband or near-broadband data rates to mobile handsets. Data capabilities were built in from the start and can be used to connect to the Internet and to remotely access the corporate network. Because coverage is currently somewhat limited, it is not a universal solution because fallback methods (such as GPRS) are needed where coverage is poor or nonexistent. Whereas UMTS supports up to 2 Mbps, and with additional capabilities a little more, fourth generation mobile communication systems already under development support 10 Mbps or more and allow parallel use by several users within the same cell, which is currently the weak point of UMTS. UMTS data rates drop with parallel usage, so the achievable speed can be unpredictable.

4.5.3 Wireless LAN

Wireless LAN hotspots are small areas where high-speed connections to the Internet are available, based on a family of standards named 802.x of the Institute of Electrical and Electronics Engineers (IEEE). On the user's side, the equipment necessary is built into almost every notebook computer available on the market today. If a WLAN-equipped notebook is within reach of a hotspot, a high-speed Internet connection can be established and VPN connections to the corporate network can be made based on this.

WLAN hotspots can be found in areas with a high frequency of people passing by, such as hotels, airport waiting areas and airline lounges, train stations, coffee shops, restaurants, gas stations, and university campuses (see Figure 4.10). Although some hotspots are privately owned and are open to the public as a courtesy service, large telecommunication companies, local ISPs, or specialized WLAN service providers provide most of them. The latter charge the users through a variety of methods, but mostly by the use of prepaid vouchers.

Because there are so many different providers and payment schemes it is not easy to use WLAN as a business tool. Business WLAN user requirements are as follows:

- Simple authentication;
- Simple and easily understandable pricing structure;
- Roaming capabilities (between different WLAN service providers nationally and internationally);
- Sufficient bandwidth;
- Compatibility of the devices used;

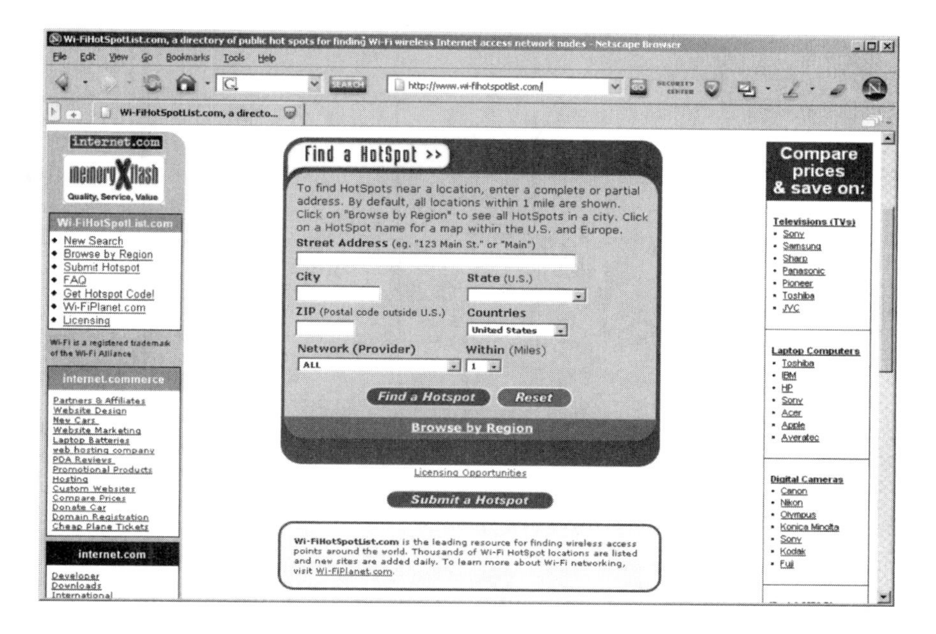

Figure 4.10 Online resource for finding hotspots around the world.

- Support for voice over WLAN (sometimes this is done by setting up a different router on another frequency, or channel, to split between voice and data traffic since QoS is currently not supported in WLAN implementations);
- Data security.

These requirements are currently only partially fulfilled.

Because WLAN is only available in selected locations and has limited range, a combination of WLAN and mobile phone-based data services would—if delivered tightly integrated—be the best tool for supporting business travelers worldwide.

Besides the public hotspots, many corporations use WLANs as a private service within their own company locations. Voice over WLAN is still a problem even in the closed user groups of corporate environments. Future implementations of the WLAN standards will be able to support QoS.

4.5.4 Mobile Data Services and "Always-On" Connections

If one examines only bandwidth developments, then one misses an important change that came with the advent of packet-oriented mobile data services, such

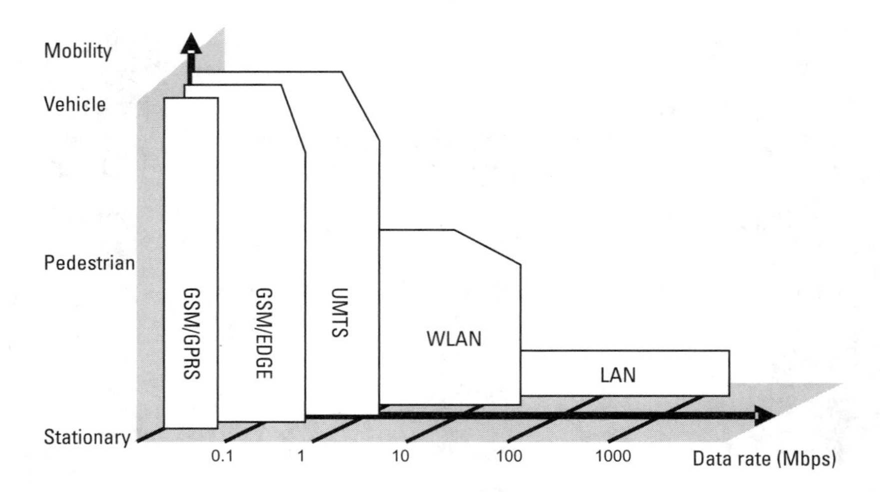

Figure 4.11 Data rates and mobility. (*After:* [8].)

as GPRS. New service offerings utilize the potential of packet data and allow "always-on" use. As described in Chapter 1, "always on" has major benefits for corporate use. Without dialing/establishing a connection, the user has instant access to corporate data and any messages that are relevant to her. Packet-oriented mobile data therefore lays the foundation for real mobile work (as long as the bandwidth is adequate).

4.5.5 Mobile Data Services and Service Integration

By combining different mobile data services, companies can provide mobile users with an environment that makes mobile work useful and rewarding (see Figure 4.11). A connection to the corporate network is then made over the best available connection. Near a hotspot this would be a WLAN, within a city probably UMTS, and in rural areas GPRS or another mobile data service provided by the mobile phone standard used.

In an ideal world, cost issues and transmission times would also be estimated in advance; for example, any e-mail with large file attachments would be placed on hold until contact with a hotspot can be made. (With GPRS this would take too long and compromise other communication needs, and with 3G data it would probably be much too expensive.) So a seamless combination of these services would not only require a consistent user interface, but also a parameter-based system to assist with steering communications by following prioritization rules, and cost parameters to decide what to transmit and what not. Service integration and roaming capabilities are the user requirements that must be addressed by the service providers and software developers.

References

[1] Yankee Group, 2004.

[2] *Telecom Dictonary*, Althos Publishing, www.althosbooks.com, 2004.

[3] "Corporate VoIP Market, 2004–2008," Radicati Group.

[4] Swyx Communications, "IP Telephony Guide," www.swyx.com, 2004

[5] "Survey," Mercer Management Consulting, 2004.

[6] MetaGroup, 2004.

[7] "IP Voice Services: The Return on Investment," Analysis, Autumn 2002.

[8] "IP Convergence in Mobile Communication Systems," *The Network Insider*, Comconsult, Aachen, Germany, May 2002.

5

Make or Buy—New Answers to an Old Question

If a seaman doesn't know what shore to steer no wind is the right one.

—*Lucius Annaeus Seneca (4 b.c. to 65 a.c.),*
Roman philosopher and poet

5.1 Self-Provisioning Versus Outsourcing

The decision between self-provisioning or buying services is repeatedly asked for different areas and departments of the enterprise. It touches a core function of doing business: how to make the best out of limited resources. This is a strategic as well as an operational issue.

Although cost issues dominate outsourcing decision making for the most part, a shortage of personnel resources as well as missing knowledge in certain areas also influence the decision-making process. Other views of the enterprise also lead to outsourcing-related discussion. The article "Unbundling the Corporation," by John Hagel and Marc Singer [1], sees the corporate entity divided into three distinct parts:

- A customer relationship business;
- A product innovation business;
- An infrastructure business.

Hagel and Singer suggest that the company should decide in which of these areas they want to compete with others, and also recommend that the other area or areas should then be "unbundled." This relatively simple concept leads the way to a strategic view of far-reaching outsourcing decisions and sees the company as an "extended corporation," a company that builds "smart alliances" (as authors John Harbison and Peter Pekar describe in their book named after this term [2]) to succeed on the global market. Technology plays a key role in this smart enterprise.

Discussions about the core competencies of corporations include debate about large-scale outsourcing. Since the article "The Core Competence of the Corporation" was published in *Harvard Business Review* in 1990 [3], the core competencies concept has become known to a wider public, is now common knowledge in strategic management, and is still being discussed in dozens of articles and books ever since. Basically, being good at your core competencies will provide your company with competitive advantages. Core competencies include every ability that:

- Is valuable;
- Is scarce (not readily available);
- Is difficult to imitate;
- Is difficult to substitute for;
- Improves the more the organization makes use of it.

A core competence can be integrated into a product but this is not always the case. Marketing and design can also be a core competence.

But even activities that most of us would see as core competencies, such as production of the goods themselves, are outsourced in several industries. So all of the leading sporting goods manufacturers do not have factories of their own and use external suppliers to produce the articles. The brand owner only does the design and marketing. The same is true for large parts of the textile and fashion industries. Most fashion houses rely solely on their brand name and design abilities and let other companies do the manufacturing. From the perspective of an industry insider, the core competencies of the textile/fashion and sporting goods industry are product development, design, and marketing—not manufacturing.

In home electronics, computer hardware, and mobile phone products, other manufacturers with brand names of their own are very often used as suppliers. This sometimes goes as far as including development services into the outsourcing contract. Only the brand management and sales remain as the areas of expertise in such cases.

Even the automotive industry outsources its manufacturing. Porsche uses the manufacturing plant of the automotive assembly corporation Valmet in

Finland as an extension of their own capacity to build the Boxster and the new Cayman model. The automotive parts supplier Magna has matured and now delivers complete assembly services to several customers such as BMW (X3 SUV) and Saab (convertible). Both cars are built exclusively at Magna plants in Austria. With more than 200,000 cars built in 2004, Magna considers itself the "world's biggest car manufacturer without its own brand" (according to the Magna press office). All of the manufacturers presented here as examples of the automotive industry still build their own engines and supply these to the assembly lines, but even this is under discussion.

When it comes to computer hardware, or mobile phones and telephony equipment in general, external suppliers do most of the manufacturing. Manufacturing within these industries can now largely be seen as the assembly of industry-standard components. Several successful companies in these areas outsource their manufacturing, but the inherent risk with limiting your own company to design and marketing is that, in the long run, the manufacturing partner may want to extend its business by creating its own brand, at the cost of the partnership. Because the manufacturer already knows everything necessary to produce the requested goods, the only market entry barrier for it is the brand. Although brand building is a costly issue, one could easily buy an abandoned brand and resurrect it. Several Asian manufacturers have bought themselves into markets in the United States and Europe using this strategy. So, crucial for success in such a potentially hazardous outsourcing relationship is for the brand owners to always be one step ahead of their manufacturers in development and design.

So how are corporate networks and IT systems core competencies of the corporation? How do outsourcing decisions affect these core competencies? Of course, in some companies, IT and TC are already core competencies. This group includes carriers and service providers, as well as IT-centered companies such as Amazon or eBay. As mentioned in Chapter 3, corporate networks and IT systems are nowadays being increasingly viewed as commodity services, such as electrical power or water supplies, rather than core competencies.

Cost issues still remain the primary driving factor of outsourcing deliberations. Per a study by Pierre Audoin Consult [4], other motives include increased efficiency (50%), increased availability (45%), and better service quality (40%). These numbers are very close to numbers found in several similar studies published during the last few years. If one goes into more detail here, there are other aspects to consider apart from cost, service, and availability. In a study titled "Update on the Outsourcing Market: Trends and Key Findings" [5], the analysis company Dataquest describes the following discussion topics as being relevant for outsourcing decision making:

- Better service quality;

- Focusing on core competencies;
- Increasing the efficiency of the IT department;
- Adding IT personnel and resources;
- Achieving a match between IT strategy and business goals;
- Gaining technical expertise and know-how;
- Improving business processes;
- Migrating to new technology platforms;
- Gaining competitive advantages;
- Gaining access to process and industry knowledge;
- Reducing costs and personnel;
- Increasing shareholder value;
- Minimizing risk.

These topics can be found in several similar lists in other publications, but with a different focus and prioritization. When asking companies who are in the middle of the outsourcing decision process why they actually do it, statements such as "We have to do something about IT" and "Things can only get better" are sometimes heard, which gives us an indication about how attempting to solve things internally has failed.

But most corporations share the same basic opinion that the question is not *if* IT/TC services should be outsourced, but *how*. Also, what parts of it should be outsourced to a service provider?

"How to do it" is the key process that needs to be addressed. Several outsourcing concepts are discussed in this chapter. Within the related subsections, one can see that spatiotemporal aspects are addressed, as well as conceptual questions and questions of business connections or ownership structures.

To clarify the terms used here and to make the differences and similarities between the listed concepts understandable and useful for the reader, it is sometimes necessary to broaden the discussion beyond the core aspects of "network consolidation." But one fact remains true: Outsourcing must to be approached sensibly because it touches on important issues in every corporation:

> "Are IT and telecommunications strategic factors within my competitive environment, or are they just a commodity for me?"
>
> "Are IT and telecommunications a core competency of our business?"
>
> "Are there any parts of our IT/TC that are strategically important?"
>
> "If we decide to outsource, is there a way back?"

In light of these questions, it is no wonder that outsourcing is seldom discussed without emotion.

5.2 Variations of Buying Services

5.2.1 Outsourcing

Outsourcing includes and integrates the different facets of buying services from external providers. The original concept of outsourcing includes not only someone else providing services, but also the movement of personnel from the company that outsources to the service provider.

From an historic point of view, the decision of Eastman Kodak back in 1989 to buy IT and communication networks as services from DEC, IBM, and Businessland instead of provisioning it with its own staff can be seen as the first (and most important) step toward the outsourcing boom now occurring within IT and telecommunications. Some authors of outsourcing books also date the roots of outsourcing in the United States to the 1960s, but this discussion is fruitless since it is simply a thing of the past. With the emergence of the Internet and the new convergent networks, a new environment has occurred that is providing chances for a renewed discussion on anything relating to outsourcing.

Public opinion is widely negative when it comes to outsourcing. The public mind recalls outsourcing deals that went bad and the related press coverage that dominated the public discussion. And even in the industry press and publications specializing in IT/TC, negative opinions sometimes dominate the scene. Several studies published view 50% or more of classic outsourcing deals as failures. The Gartner Group mentioned in a recent report [6] that 85% of all outsourcing contracts that were underwritten since 2001 had to be adapted and the contract reworked, since the contracts were longer suitable for supporting the long-term company goals.

The management consulting company Diamond Cluster discovered in 2002 [7] that more than 70% of all U.S. companies asked to end an outsourcing deal earlier than stated in the contract. The reasons were (1) service problems that occurred, (2) costs that were too high, and (3) changes in company strategy. In 71% of these problem cases, a different provider was chosen, while in 29% the company decided to bring the service back in-house.

The capital market has no strong opinion on outsourcing decisions in general. A study by E-Finance Labs, the German competence center of the financial service and banking industry of Frankfurt University and Darmstadt Technical University [8], came to the conclusion that outsourcing itself has no positive effects on the share price of a company within the financial sector. Although this study was focused solely on the worldwide financial sector, one can still draw conclusions about the neutral effect this probably also has in other industries.

A very different view on outsourcing can be expected from any individual who is affected by an outsourcing decision, such as an employee of the company that decides to outsource. His view will be dominated by fear. He will fear the loss of his job and, if he is in the management layer, possibly the loss of influence and power.

Apart from the negative discussion and critical press coverage, the term *outsourcing* is still the most important term used and covers several ways of buying services provided by an external company. Although the original concept includes personnel being relocated from the company that outsources to the service provider, most current implementations do not include any personnel relocation.

The term *outsourcing* itself is composed from the words *outside, resource,* and *using* or *outside* and *resourcing.* Both combinations describe a move from internal service provisioning to external services. Outsourcing in the original meaning of the word focuses solely on provision of services by external independent contract partners. In this case, a contract defines the rights and duties of every party. If one discusses the legal aspects, however, a fine distinction can be made between outsourcing to external suppliers and "internal outsourcing," which includes cutting of services and moving them to a separate unit, department, or company while still maintaining ownership and control. When it comes to internal outsourcing, the partner is an internal unit or a company under the ownership of the company buying the service. Apart from the contractual rules, in this case it is possible to influence the service provider directly.

When discussing the advantages and disadvantages of information technology outsourcing, there are strategic, outcome-oriented, cost-oriented, and personnel-oriented aspects to be noted, as listed in Table 5.1.

Different types of outsourcing have different advantages and disadvantages. The next few sections take a closer look at the variations on the outsourcing theme.

Complete Versus Selective Outsourcing

One speaks of "total outsourcing" or "complete outsourcing" when all IT functions relating to desktop computers, servers, and the network infrastructure are given to an external service provider. In most cases, this also includes the relocation of personnel from the company that is outsourcing its IT functions to the service provider.

In addition to this traditional understanding of outsourcing the concept of "selective outsourcing" has gained ground during the last few years. When it comes to selective outsourcing, only certain IT functions (and also other corporate functions) are addressed, for example:

Table 5.1
Advantages and Disadvantages of IT Outsourcing

	Advantages	Disadvantages
Strategic	Focusing on the core business	Permanent dependence on the supplier
	Increasing flexibility	Lack of acceptance by the departments
	Transfer of risk	Increased coordination needs
	Standardization (maybe an increase in quality)	Cooperation risks (security risks; financial risks, for example, provider bankruptcy)
Outcome oriented	Possible know-how gain	Possible know-how loss
	Clearly defined responsibilities	Quality risks
	Service orientation	Loss of control
	Easy access to additional capacity when needed	
Cost oriented	Reduction of total cost of ownership	Transaction costs
	Cost transparency	Change costs
	Variabilization of costs	Cost allocation base is difficult to find
Personnel oriented	Reduction of recruiting and qualification problems	Personnel problems during the transition
	Making better use of own personnel resources by surrendering routine tasks	Problems with employee motivation
	Release from the need to provide multiple shifts within infrastructure operations	Difficult monitoring

Source: [9].

- Data center outsourcing;
- On-site infrastructure outsourcing;
- E-business outsourcing and e-business infrastructure outsourcing;
- WAN outsourcing;
- Business process outsourcing;
- Application service providing/provisioning;
- Application hosting/server hosting;
- Systems management outsourcing;
- User help desk.

Note that most terms listed here lack a precise definition and can have a different meaning when used in a different context. Because vendors influence the entire market, the concrete meaning of a term used is often closely related to provider service offerings. The meanings also differ from publication to publication.

Outsourcing Typology

If one wants to differentiate the term *outsourcing* without using too much "provider speech," a basic concept (first laid down by Riedl [10]) divides the term into three possible fields: (1) business outsourcing, (2) application outsourcing, and (3) technology and infrastructure outsourcing. These three basic terms can be further divided into individual solutions and standard, off-the-shelf solutions.

Business outsourcing includes business process outsourcing as an individual service, which is described in an upcoming section, as well as business processing as an off-the-shelf solution, which is quite popular when it comes to, for example, standardized banking transactions.

Further distinctions can be made between application hosting as a service provided individually and application service providing/provisioning (ASP) as a standard solution. When it comes to technology and infrastructure outsourcing, a distinction can be made between off-the-shelf solutions, such as basic Web hosting, and individual agreements relating to outsourcing the existing corporate network infrastructure. Although this differentiation cannot be seen as complete or final, it helps us gain a better view of the outsourcing area by accounting for the various degrees of individualization.

On a list published in 2004 [11], the center for outsourcing studies at St. Gallen University in Switzerland named more than 30 terms that can be seen as part of the master term *outsourcing*. Depending on the vendor or publication, the use of any of these terms varies. "ASP" services provided by service provider X do not necessarily share any of the characteristics of "ASP" services provided by service provider Y.

Other factors should also be considered, especially if one wants to gain knowledge for one's own business, by analyzing large public outsourcing deals. Payments for other services are very often hidden within the sums of money mentioned in the news. In particular, costs for one-time services, such as systems integration, are very easily hidden in long-term running outsourcing deals. Apart from the sums of money necessary for monthly servicing and other tasks, such as software development, ERP implementations are sometimes priced into a single contract and lead to monthly payments instead of one-time fees.

5.2.2 Offshoring

Offshoring is the type of outsourcing that is most discussed in public. Offshoring describes the outsourcing of company activities, tasks, and business processes to countries with lower wages to bring down total costs for IT development, system maintenance, and infrastructure administration. The main reasons for offshoring are the differences in wages between different countries. Other driving factors include increased quality and acceleration of processes [12].

So what activities are best suited for offshoring? The market analysis company Datamonitor sees the activities listed below as the most relevant [13]:

- Activities with a low necessary knowledge profile but high intensity, such as typing data in large volumes. These activities build the roots of offshoring. Doing simple tasks with cheap labor from remote countries has been happening for decades now.
- Tasks with medium requirements, which can be done using a set of guidelines and rules and which are labor intensive. These include call centers and customer service. To provide these services in a country with low wages, the employees must have sufficient language skills (in the language of the company for which they work). A cost-efficient and well-functioning telecommunications infrastructure is also necessary.
- Tasks with high requirements that are labor-intensive. These include foremost software development and maintenance. Due to the good university education and increased skill sets within offshoring countries such as India, this segment has grown lately. These tasks are the core of the current public discussion about offshore outsourcing.

By means of remote management systems, the chances are that offshore service providers can provide many of the services that do not (or no longer) require an on-site presence. These include:

- Application development and maintenance (as listed earlier by Datamonitor [13]);
- IT infrastructure management, which itself includes system administration, network administration, database administration, security management, desktop management, technical support, and disaster recovery services;
- Business process outsourcing, such as human resources management, order entry, archiving, and call center services (a detailed description can be found in Section 5.2.4).

Offshoring and Outsourcing

Not all offshoring activities are also outsourcing activities in the pure meaning of the word, because many larger companies use offshoring as a way of building up their own capacities in countries with low labor costs. But the consequences for the employees are the same, since anyone from a developed country will probably never move to a low labor cost country when their job moves there.

Offshoring Advantages

The advantages typically mentioned include the following:

- Cost savings;
- Fast and easy scalability and flexible cost structure ("pay for what you need");
- Reduced "time to market";
- Higher availability of skilled employees;
- Better predictability and transparency due to defined processes, which can lead to better quality;
- Cutting off of noncore activities.

Of course, the differences in wages are the main driving factor for outsourcing in all of these items.

Offshoring Risks and Disadvantages

Many companies are so keen on cost savings due to reduced labor costs that the risks and disadvantages that can eventually kill entire projects are denied or simply not seen. Offshoring risks and disadvantages include the following:

- *No savings achieved.* The buyer of services is confronted by extra costs in the total cost calculation due to hidden costs that were not seen when planning offshoring (because, for instance, the buyer of the service did not verify the vendor's optimistic views).
- *Loss of knowledge.*
- *Language barriers.* Although most personnel within offshoring companies have English language skills, one should not expect that communication with them will be as easy as with native speakers. Problems due to misunderstanding must be expected.
- *High costs during migration period.* For moving services offshore, one must expect a time period of about 3 to 12 months during which additional costs may occur. These migration-related costs are often

underestimated. Even after migration has ended, some of these cost components remain: costs for project management, international (sometimes intercontinental) travel, and communications costs.

- *Cultural problems.* Especially when it comes to offshoring to Asian countries, cultural differences can lead to problems. The Asian mentality of consensus can lead to problems occurring in the later stages of a project.
- *Time zone differences.* This is not a problem when outsourcing to Mexico, but is potentially a major problem when it comes to India or China. But this could be helpful when a company is merely offshoring additional capacity to these countries, in establishing "follow-the-sun" 24/7 activities.
- *Distance.* Distance can be a real problem since personal relationships cannot be established, or established only to a limited degree, when people each other only via the phone and e-mail. Larger Indian offshore companies address this problem by placing service units nearer the customer within the United States and Europe. They sometimes call this additional service "nearshore."
- *Breakdowns at the service provider.* This carries a huge risk since one has no access and no control.
- *Communication infrastructure breakdown.* Although this rarely happens, the risk is higher when offshoring than with outsourcing to a service provider in your own country.
- *Political and economic issues in offshore countries.* Depending on the country, potential problems such as war or power supply problems could arise. Developing countries are especially at risk.
- *Side activities and cheating.* This is a topic nobody talks about loudly. With software development in particular, there is a large risk that the offshore service provider will gain significant knowledge from their software development contracts. It is not improbable that this knowledge gain will include the use of your company's source code in other products; for instance, in similar products of other companies who are also buying software development from the provider, or in products that the providers bring to market themselves. In both cases, the competitive advantage is compromised and direct financial losses can result. These are real problems seen by anyone who looks closely at the Web sites or brochures of offshore service providers. In these, one can very often find software product offerings that are similar to products already on the market. When it comes to service outsourcing, however, the risk of this happening is limited.

- *Problems with privacy.* Regulatory issues and laws regarding privacy of personal data differ from country to country. This can be a problem when offshoring, since countries such as India have no or few privacy protection laws.
- *Problems with data security.* One will not know what the outsourcing company does with the data it works with. This also includes the risk of corporate espionage.
- *Scope creep.* This is not only a problem relevant to offshore business. Changing requirements lead to additional costs. The increase can be steeper when offshoring.
- *Changes in key personnel.* Typical offshore countries such as those located in India are experiencing an economic boom. Employees changing employers is the norm and cannot be influenced by the outsourcing company.

These risks cannot be ignored. Even large Indian service providers see a necessity to provide information on the risks of offshoring within their own sales presentations, and even include an estimation of a percentage of projects that fail.

A study by the Gordon & Glickson law firm recently came to the conclusion that only 38% of all offshoring projects can be seen as successful, 27% as at least partly successful, and 35% as failures [14]. Within their publication, they list a number of reasons for the failure of offshore projects:

- Problems with maintenance and support;
- Higher costs than expected;
- Misunderstanding by the service provider of the core business;
- Missing reactions on requirement changes;
- Failure to build up a good relationship;
- Loss of management control;
- Underestimation of the technical deployment workload;
- Important delivery dates that were not met;
- Loss of key personnel.

To sum this up, one can say that offshoring is a high-risk business and one thing is true in the long run: The economic development in the country providing these services will eventually kill the cost advantages due to higher labor costs. For instance, Ireland, a country that was once was a favorite place for providing call center services, is already past the peak of providing offshoring services and is now steeply declining. Due to increased wages, companies looking for cheap labor have already moved elsewhere.

Implications for Society

The public offshore debate is dominated by the fear of losing work within the homeland. Current examples show that almost every major U.S. company has at least some work that has been moved abroad. "My job has gone to India" is a sentence often heard. So it is no wonder that there are companies out there who want to profit from this, such as the IT service company MPC, which began a "Buy American" campaign in 2004 (see Figure 5.1).

One would expect that offshore outsourcing would lead to higher unemployment rates, at least in IT and TC. Some studies, however, show that the effects on the number of workers in the United States or in other countries that buy offshore services are *not* negative. A study by the market research company Global Insight [15] came to the conclusion that, in 2003, 104,000 IT occupations were lost due to outsourcing (including but not limited to offshoring), which due to productivity increases led to 193,000 new jobs in other industries within the United States. Global Insight concludes that this is indeed a total increase of almost 90,000 jobs, instead of the expected decrease.

Figure 5.1 U.S. company Web site of MPC Computers.

There is an interesting annotation when it comes to start-up companies. *Wired* magazine described in July 2004 [16] that venture-capital firms often urge start-up companies to outsource all their development work. *Wired* magazine went on to say that in Silicon Valley they already have a name for it: *micro-multinational.*

Nearshore Outsourcing

Within Western Europe, the term *nearshore outsourcing* is used as a description of sourcing out services to nearby countries (such as Poland to Germany). Indian offshoring companies also use this term but with a different meaning: To them, *nearshore* means providing contact personnel within the country where the company that outsources resides.

5.2.3 Application Service Providing and Managed Services

Delivering software applications as a service was one of the revolutionary ideas and developments of the "e-business hype." Several start-up companies tried to fulfill the ASP promise. At this time, ASP was seen as the software delivery concept that would become the most important business model for software. Renting or leasing applications, instead of buying, was expected to dominate the market. Even simple standard applications such as Word or Excel were no longer to be installed and used on the client's computer, but instead delivered on demand via the provider network.

Most start-up companies in this market have already left the scene. Other vendors, such as Internet service providers that also delivered ASP services, have also changed their business model or stopped offering it.

Although these first attempts at ASP did not succeed, there are undeniable advantages:

- Applications can be built according to usage. If certain applications are seldom used, then one can expect major cost advantages.
- Software is never out of date. Because any new release is instantly available, and patch management is centralized, there is no worry about security holes.
- Access to software applications is not bound to any location but is available everywhere (as required).
- Operation of the software is done in the provider data center with automatic backup and disaster recovery precautions.

Another advantage sometimes mentioned is that up front investments in software are low and, if a project fails, the software used for it can be canceled. If the company had bought the software instead of renting, these costs would remain since the chances of reselling the once bought but no longer used license are low.

Whereas first generation ASP services focused on providing well-known applications in a different way, there are also software concepts that focus on ASP in the foreground, such as Web content management systems or e-commerce software platforms.

Hosted Services

Although ASP in general had (and still has) problems gaining market share despite the hype in the 1990s, external provision of Web-based applications is growing. Administration and maintenance of any Web application that goes beyond HTML Web pages, such as a complex corporate Web site or an online order system, leads to high customer expectations of availability. Web users do not usually conform to service center working hours but use the system 24/7. Server computing power and bandwidth are not easy to estimate because of usage peaks. This is much easier when using a data center to provide several similar services rather than just providing one. Because peaks do not usually occur over the same time period on several systems simultaneously, the average bandwidth needs per unit will be lower. From a cost as well as an availability point of view, providing a Web service externally has advantages over self-provisioning.

Industry analyst companies such as Gartner expect an increase in hosting revenues for service providers, despite the price cuts of the past. The market for hosting is fragmented. Several hundreds, and maybe thousands, of companies provide hosting or hosted services. Since many of the vendors currently on the market are just reselling other providers' hosting capacity, the real number of service providers is much lower than expected. Within these, even fewer offer more than simple hosting, for example, provision of online ordering systems or content management applications.

The term *hosted services* is not always used in the same way. *Hosted services* is sometimes just used as a synonym for ASP. A broad use of the term can include Web site hosting, hosting of messaging software, telephony services, administration of desktop PCs, back office applications (for example, calculation of salaries), transaction systems (such as business-to-business online ordering), and ERP application hosting (see Figure 5.2). *Hosted services* is sometimes seen as a developing issue, from the simple services of the past to the more complex systems of the future.

E-Mail Hosting

Smaller companies are targeted by hosted messaging services as a variant of hosted services. These services are very often bundled with Web hosting as a more or less free option. The service is provided in the data center of the

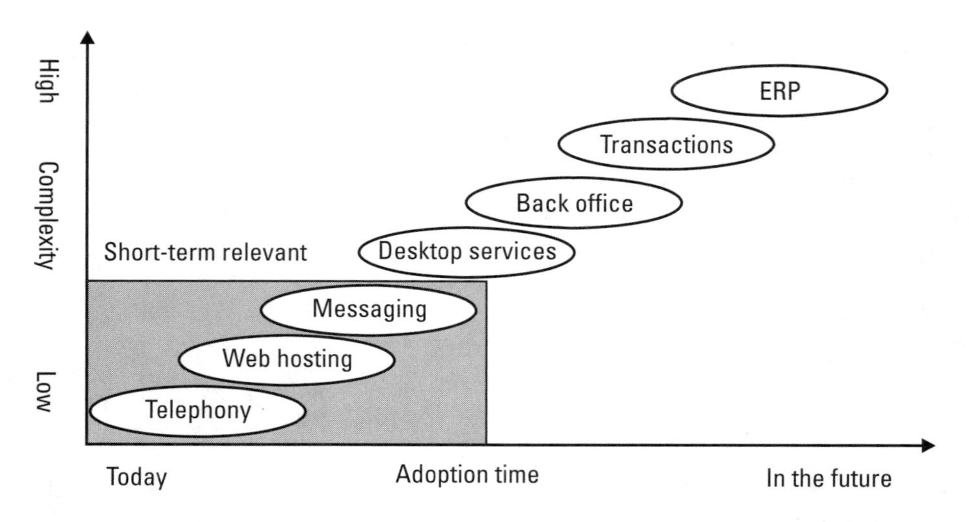

Figure 5.2 Evolution of hosted services. (*After:* Mercer Management Consulting, 2004.)

provider where the Web hosting is also located. Newer solutions integrate elaborate messaging applications, for example, providing Microsoft Exchange functionality or offering other groupware functionality to small companies. Of course, these functionalities are then priced separately, while the basic messaging functions remain free.

Industry analysts expect the hosted e-mail market to grow steadily during the next few years. The public discussion on e-mail security and spam problems will influence the decisions of companies as to whether or not they will outsource their messaging services.

It is up to the reader if she now chooses to see Web hosting and e-mail hosting as a kind of ASP service or prefers to use the term *managed hosting* as a higher-level concept. A consistent description cannot currently be found in the literature.

ERP Hosting

Providing ERP systems as a service can be seen as a separate domain, but also as part of the ASP service model. It is expected that ERP hosting will increase, especially since more and more smaller companies are investing in ERP and do not want the large capital costs associated with self-provisioning.

Managed Services

One can say that managed services are a variant, or a further development, of ASP. Managed services provide additional services such as security management, backup, or database optimization besides simple provisioning of software, as in the original ASP model. These services are priced according to usage or use a time-based flat fee (for example, a monthly fee). Although this model has undeniable advantages for companies using these services, and also provides chances of a continuous revenue stream for providers, scalability of services remains questionable.

The term *managed services* is also used differently within the literature and vendor marketing material. Managed services focused on infrastructure, *managed network services* are therefore separately treated in a specific chapter later in this book.

5.2.4 Business Process Outsourcing

Business process outsourcing (BPO) in general stands for the outsourcing of complete business processes that are not within the core competencies of a corporation. Examples of business processes that can be easily outsourced while not touching the core competencies are payroll accounting, billing and bill collection, and procurement of standard parts. BPO often includes IT.

The idea behind BPO is not new. The concept first appeared in the middle of the 20th century, as companies gave company functions from the financing, logistics, law, and accounting departments to external service providers. Of course, it was not called BPO back then. The name is just an invention of the recent past.

Although BPO has a long tradition, the term is used inconsistently. Most descriptions of BPO come from the marketing papers of service providers. Like *VPN* discussed earlier, vendor-driven definitions have led to imprecise use of the term BPO. The analysis company IDC precisely describes BPO as follows [17]:

> BPO involves the transfer of management, and execution of one or more complete business processes or entire functions, to an external service provider. The BPO vendor is part of the decision-making structure surrounding the outsourced processes and functional area, and performance metrics are primarily tied to customer service and strategic business value. Strategic business value is recognized through results such as increased productivity, new business opportunities, new revenue generation, cost reduction, business transformation, and/or the improvement of shareholders' value.

BPO therefore has two prominent features:

- Orientation to strategic values (for example, increasing service quality, increasing shareholder value, opening new markets);
- Individuality within the business relationship (customized services delivered in a "one-to-one" approach).

A very common example for BPO activities is call center outsourcing to a specialized service provider. In Table 5.2, the analysis company Yankee Group delivers a broad overview on business processes that can be outsourced to an external service provider [18].

We can make a further distinction between BPO and "processing services." The difference between BPO and processing services is the degree of individuality that comes with each service. While processing services focus on standard processes done everywhere in the same way (such as payroll accounting), BPO applies to individual services, such as those needed for, say, recruiting services. This distinction also highlights the boundaries of what makes sense and what does not when it comes to business process outsourcing. Most companies would not surrender the recruitment of management and highly skilled workers to an external service provider, but will rely on recruitment services for normal blue-collar workers.

5.2.5 Outtasking

A very selective outsourcing concept is described by the term *outtasking*. The term *selective outsourcing* is also sometimes used for the same process. *Outtasking* refers to buying single tasks, which are usually done within the company, from an external service provider as a part of supporting the core business. For example, the operation of a company switchboard or a WAN can be such a task. When it comes to corporate telephony in particular, outtasking has been done for years, although it has only been named such in recent years.

Table 5.2
Taxonomy of Internal Business Processes

Production	Research and development, production, maintenance, packaging
Sales, marketing, communication	CRM, marketing, call center, external communications (press, public relations)
IT and telecommunications	Desktop management, network and server, management and maintenance of applications, telecommunications
Human resources	Training, recruiting, payroll accounting, personnel administration
Supply chain management	Storage, transport, logistics, fleet services
Administration and financial services	Legal advice and tax accountancy, insurance, controlling, accounting, billing

With network outtasking, an external service provider provisions, operates, and maintains the corporate network (WAN, LAN, or both). The company either retains ownership of the network devices or transfers this to the servicing company, depending on the contractual situation.

5.2.6 Other Types of Outsourcing

On-Demand and Utility Computing

Much public discussion occurs in relation to provision of "on-demand" services, especially on the topic of "on-demand computing." Marketing drives this concept, of course, but it sounds simple and appealing to potential customers. In reality, service delivery is much more complicated than expected.

On-demand computing occurs when computing resources are offered and accepted in a demand-oriented way and billing follows the actual usage. Within this context, the term *utility computing* is also known, but the similarities between both concepts are limited. *Utility computing* refers to complete outsourcing by an ASP model and managed services.

On-demand computing has a certain appeal to most of the people responsible for IT within corporations. The idea that, by merely pushing the right button, one can get all of the necessary computing services and at the end of the month have each of those services listed precisely on a single bill sounds to them like the holy grail of computing management. The current reality, however, looks very different than these expectations, which have largely been generated by marketing hype. Existing infrastructure components, as well as strategic discussions on how IT can be a weapon in competition, hinder the fast adoption of this idea in corporate reality. As in any other outsourcing-related topic, cost issues play the major role, while security and performance guarantees are also important when it comes to buying on-demand services.

Almost every ERP vendor now offers some kind of on-demand service or at least a usage-based pricing model. The relation to ASP services is so close that distinctions between those two models are sometimes hard to make. For the future success of on-demand services, the need for clear definitions and a common understanding of the methodology must be addressed.

Business Transformation Outsourcing

Business transformation outsourcing (BTO) is a typical outsourcing-related term, originating from a consulting company. Similar to the BPO concept described earlier, a service provider takes over a business process from an outsourcing company. While this may come with or without changes in BPO, the BTO concept changes this process and—if necessary—completely redesigns it.

BTO can be used in all areas where one would expect efficiency gains due to process improvements, for example, within the areas of:

- Finance and accounting;
- Human resources;
- Procurement;
- Logistics;
- Customer relationship management;
- Technical services;
- Customer and employee support.

The goals of BTO are better processes and cost reductions. To successfully provide BTO services, the service provider needs to have industry knowledge, as well as a deep knowledge of the business processes affected. The service provider assumes complete responsibility for the business processes affected.

For the success of an outsourcing concept largely based on transformation concepts, flexible contract articles are the most important feature. The sharing of risk and success between the partners should be clearly defined. Only when the contract allows the service provider to participate from efficiency gains will it be interested in gaining greater benefits than those to which they are contractually bound and originally expect. BTO contracts are typically medium and long-term relationships covering a time period of about 5 to 10 years.

Describing a strict distinction between BTO and BPO is hardly possible. BPO is not limited to taking over processes and, in most cases, also attempts to provide efficiency gains. BTO can therefore be seen as a clarification of what was originally intended with business process outsourcing.

Precision Sourcing

Another buzzword within the outsourcing discussion is *precision sourcing*. This dates back to publications of the Yankee Group, which may not have invented the term itself but have certainly brought it into the discussion on outsourcing. Yankee sees precision sourcing as a type of selective outsourcing. In summary, one can say that it is similar to the outtasking concept described earlier.

Managed Network Services

One speaks of *managed network services* when not only the network connection, but also the related equipment within the customer's site is provided and serviced by the network provider. In this case, the service provider assumes all responsibility for service delivery. This usually also includes support services.

5.3 Buying Voice and Data Networks as a Service

With a broader understanding of what outsourcing is, and what it can include, evolving during the last few years, the term *managed services* now plays a major role when it comes to outsourcing corporate networks and network-related services. Managed services includes operation, maintenance, and support for company networks but without necessarily taking over assets and personnel. Managed services can also focus on single network services, for example, firewall management or messaging services.

Apart from the usual cost discussions, several benefits accrue from outsourcing network-related services such as network management, firewall management, help desk, provisioning, and operation of network-related applications:

- Availability of 24/7 services, instead of the usual one- or two-shift work organization provided by typical companies.
- Economies of scale at the service provider, due to parallel management of several similar networks in a single network operation center. The cost savings achieved by this can lead to lower service fees.
- Increased security, since the provider can invest in highly trained, skilled specialists (for example, for security management) that an individual company usually cannot afford.

While these advantages are true for all network and network-related services, further benefits are to be found within certain areas of managed services. For instance, when it comes to WANs, a provider can deliver what the customer really needs: a defined "end-to-end" quality, by offering a provider-based IP-VPN instead of the integration of single connections that the company would usually use when self-provisioning a WAN or a public Internet-based VPN.

Within the company LAN, the situation is different. Although a significant percentage of all medium and large companies have already outsourced their desktop and server systems to a service provider, corporate LANs have very often been built in a complex way because of unstructured development that occurred across several years. Continuous changes and extensions are done almost every day. Most companies fear that service providers will not be able to deliver the needed flexibility when it comes to LAN management or that this will be too expensive.

5.3.1 Voice Data Integration and Outsourcing

As described earlier, IP voice concepts are attractive to companies. On the one hand, one gains new application possibilities, while on the other, total costs are lower when compared to traditional provisioning methods. Besides this, VoIP implementations can also be delivered as a hosted service and even allow the removal of costly company switchboards. Because PABX/PBX usually comes with a long-term service contract, an immediate switchover to VoIP is very often not economically wise.

Another aspect of the convergence development discussed in earlier chapters—the integration between mobile telephony and fixed-line telephony— will lead to additional positive effects in the long run and will also be addressed by service providers. Rethinking corporate voice communication under these preconditions is very often the most important step in outsourcing voice communications, or even the whole voice-data infrastructure, to an external service provider.

5.3.2 Web Applications and Outsourcing

A major consequence of the Internet revolution was the ubiquitous spread of Web applications. Almost all companies in business today have corporate Web sites. While smaller companies tend to buy hosting services from specialized Web hosting providers (which led to the development of a new industry in the 1990s with millions of customers worldwide), large companies act differently. Several large companies have moved their Web activities from self-provisioning to outsourcing and then back. Sometimes this process has happened more than once. This seems to be strange behavior from an outsider perspective but can be understood when considering the changes in significance that the companies have seen within their Web activities.

In the mid-90s, many companies were just experimenting with the new medium. Strong business ties, such as online order systems, made in-house provisioning seem necessary. Subsequent better provider offerings, and problems with self-provided Web services (lack of 24/7 support, unexpected traffic peaks), have led to more outsourcing relationships.

When total cost of ownership is used as a decision factor, most companies come to the conclusion that outsourcing is the way to go, since server hardware and software, database licenses, content management systems, and the necessary maintenance and support are usually more expensive then buying these as a service offering from a specialized provider who can profit from economies of scale.

5.3.3 Messaging Systems and Outsourcing

Similar to Web applications, messaging systems are very often outsourced and can be bought as an ASP service as described earlier. Basic messaging is sometimes even a free option when buying Web-hosting services. But because large parts of e-mail communication occur within the company, and not between the company and the outside world, it is not always useful and cost efficient to buy these services from an external service provider with e-mail hosting in their data center. So most large enterprises provide messaging services by themselves. Basic hosted messaging services also lack the type of functionality needed by large enterprises, such as centralized phone books, groupware functionality, and integration of fax and voicemail. The price of this is that companies suffer under today's e-mail problems, such as virus and spam protection, and the necessity for continuous patch management.

One must expect that, with an increased tendency to outsource company networks in general, outsourced messaging services will also be in demand, as a complex feature of the WAN/LAN management and no longer just as a simple Web-hosting option.

5.3.4 Complete Telecommunications Outsourcing

Single tasks and single processes are very often considered for outsourcing, but most companies are not yet thinking about complete outsourcing of the entire voice and data infrastructure and related services. Most companies fear a loss of control when it comes to complete outsourcing. But there are significant benefits to doing so that must be considered:

- Process improvements and efficiency gains compared to self-provisioning can lead to cost savings.
- Consolidation of a number of services and contract relationships within a single contract can lead to a simplified contract and fault management.

Complete Telecommunications Outsourcing with Only One Provider Worldwide

Outsourcing all corporate network activities to only one service provider worldwide is a major challenge but can be rewarding:

- A single SLA can be defined for the whole network and is usually offered by the service provider, although regional or application-specific differentiation can be a better strategy in some cases.
- There is only a single point of contact for all service requirements. Because there are no problems with finding the vendor who is responsible for a certain problem, single sourcing will save time and costs.
- Other network-related services such as messaging can very often be easily added if the basic infrastructure is already outsourced to one provider.

On a global scale, single sourcing will not work in every scenario since no single service provider truly covers all continents with the necessary fine network infrastructure.

When looking at global companies, the location of the data centers is very often an important factor in network consolidation. Nowadays it would easily be possible to consolidate all services and applications needed worldwide into a single data center. Bandwidth and QoS should not be a problem with current technology. A typical global company will have a data center within every target area, or every target continent, in which it works. A service provider acting as a "one-stop-shop" must provide these services within these areas of activity and therefore must rely on a strong and wide-reaching infrastructure of its own. Especially when it comes to mergers and acquisitions, total outsourcing can be the easiest way to integrate and consolidate different network infrastructures.

To address the problem mentioned earlier, that probably no single provider will be able to provide the same high-quality service worldwide, a single provider can of course team up with local players and act as the prime contractor who steers the others. This will also lead to better support in regions that have special requirements for service delivery.

Complete Telecommunications Outsourcing with Regional Providers

When it comes to complete outsourcing, a regional concept can be the right strategy to fulfill a company's needs. In this concept, there will be one prime partner in every region where the company is active. At first look, this strategy, which is sometimes called "federal," has certain advantages. It will be cost effective, since there are several providers with regional coverage available that will make bargaining easier. It is also possible to define regionally specific SLAs. This is important when it comes to service hours that are not 24/7 but must be adjusted to the local needs (within that time zone).

But there are also disadvantages with this regional concept. The first problem is largely related to the interfaces between the different providers and the resulting complexity of the infrastructure. Consistent quality will probably not be achieved over the entire network.

But consulting companies very often see "regional single sourcing" as an important strategy for buying telecommunications services in terms of cost and service quality. The Compass Group consulting company sees 20% to 50% cost savings through the use of regional sourcing [19]. For large multinational companies with a fragmented infrastructure within each country (for example, financial services and banking, retail), one would recommend a federal sourcing strategy with inclusion of large and wide-reaching networks of the major carriers within each country. With this concept, one can best use the strengths of each of the carriers within their home country.

Single Sourcing Versus Multiple Sourcing

What are the advantages and disadvantages of multiple-vendor strategies compared to single-vendor strategies when it comes to buying telecommunication services? In most cases, single sourcing in buying technology and services is more cost effective than buying the same services from two or more vendors, since concentrating the purchasing on one supplier usually leads to a better price. Contract management, support costs, and complexity are also reduced. Single sourcing also is a necessary precondition of total system performance measurement.

The downside of single sourcing is increased risk: The supplier can fail to provide the requested services in the expected quality. This will lead to change costs. When selecting two suppliers, one can limit this risk or eliminate it and can achieve additional savings by continuously bargaining with the two different suppliers.

Either single sourcing or dual sourcing can be more suitable, depending on the differentiation of services. When it comes to buying standardized offerings, so-called commodities, a dual or multiple-supplier relationship can be the best option, since flexibility in switching from one supplier to the other protects against any supplier failure and secures a high service quality, while the price can also be optimized. Services such as basic voice communication fall into this category.

When it comes to buying more differentiated and individualized services, a single-sourcing strategy can be more suitable, since demanding and individual requirements with respect to functionality, platform, and integration services can best be met by a single supplier (see Figure 5.3). Services such as VPN offerings and specialized network services fall into this category.

5.3.5 Development of the Service Provider Market

When it comes to complete or partial outsourcing of network services, one must take a closer look at the development of the provider market. Market conditions and provider offerings are constantly changing all over the world, but not all at the same speed.

Although the mantra of the whole service provider industry still circles around savings and cost reductions, vendors are increasingly investing in service offerings based on new technology. Market consolidation is occurring right now. Several vendors are already (or again) in financial trouble. Regulatory issues also threaten the business model of carriers and service providers worldwide, especially those that were previously national public carriers. Knowing this, one should closely examine the technical, as well as the financial, circumstances of any provider being considered as a source for services.

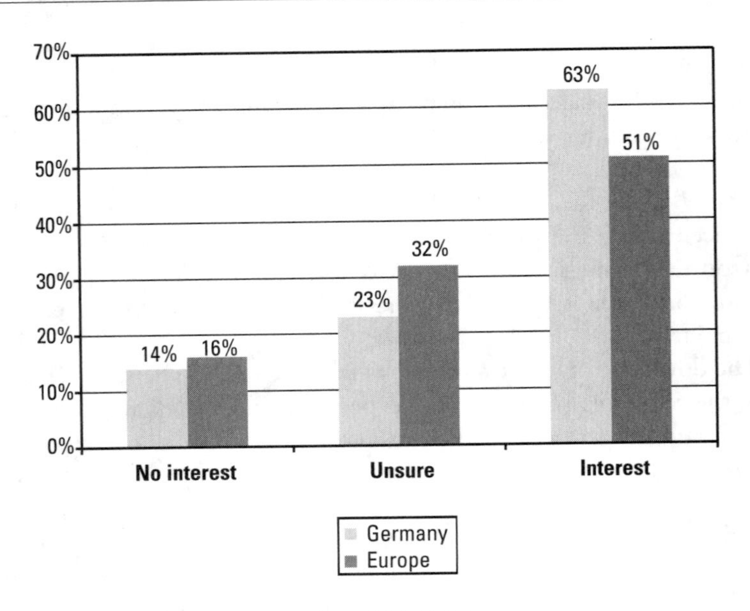

Figure 5.3 Interest in single sourcing. (*From:* IDC, 2004.)

5.3.6 Questionnaire for Buying Managed Network Services

The American consulting company Infotech provides a questionnaire for enterprises [20]. With help of these questions, every company can discover whether or not they should be buying managed network services from an external service provider:

Do you need to bring down operation costs?

Does your network infrastructure tend to converge?

Is your business success threatened when communication networks do not work?

Is there a lack of expertise for provisioning and operation of high-quality networks and network-based applications?

Is it necessary for you to shift your internal resources to your core business?

Do you experience problems in recruiting skilled employees in certain geographical areas?

Infotech concludes that, if one or more of these questions can be answered with "yes," this is a clear indicator for the necessity of buying external services.

5.3.7 Trends in Buying Managed Network Services

If one takes a closer look at the developments happening in network infrastructure outsourcing services, three basic concepts arise from different motives, as listed in Table 5.3.

5.4 Financial Considerations

5.4.1 Outsourcing and Target Costing

Target costing is a well-known concept when it comes to outsourcing. Parts of the outsourcing contract are sometimes directly related to cost targets. Within target costing, the target is simple: a percent reduction of costs from the previously calculated level. These cost savings are to be achieved through outsourcing efficiency gains.

An alternative concept sees the cost target as a fraction of company revenue, for example, IT costs must not be more than 5% of revenue. This can also be fixed in an outsourcing contract. Typical concepts include financial sanctions, as well as additional payments, when underachieving or overachieving the cost target.

To use target costing successfully within outsourcing contracts, the company must first take great care in calculating the costs before outsourcing. Critical cost components are very often missing or are difficult to identify but, without this, target costing will not be useful as a basis for defining targets in outsourcing contracts.

Table 5.3
Specifications and Reasons for Outsourcing of Infrastructure and Related Services

Specification	Reason
Outtasking	Reduction of complexity
	Filling up missing resources
	Risk minimization
Complete outsourcing	Limited know-how within the corporation
	Competitive surrounding
Using hosted services	Time to market
	Minimizing of capital cost
	Risk limitation

5.4.2 Outsourcing and Total Cost of Ownership

Even though target-costing contracts have lost importance for outsourcing decision making, a precise knowledge of internal costs is important for all contract bargaining. To provide valid data, the total cost must be calculated. Cost calculations can be very different, depending on the type of outsourcing, for example, "business process outsourcing" versus "network infrastructure outsourcing." When discussing BPO, the cost calculations very often lead to the concept of "baselining." Network infrastructure outsourcing is mostly seen as an investment with up-front costs and operational costs that are calculated in a total cost of ownership calculation.

The view on cost is changing with the wide availability of IT infrastructure and the ongoing trend toward decentralization. The initial cost (of buying hardware and software, as well as implementation work) is no longer seen as the only important component, but is now viewed as only part of the total costs related to IT or infrastructure.

Total cost of ownership (TCO) is a complete view of all budgeted and nonbudgeted costs of any investment in IT and TC, over the complete life cycle of the investment. This concept was introduced with regard to desktop PCs, by the Gartner Group analysis company in 1997, and has been used for other components of IT and telecommunications systems ever since. Almost everything IT related is already under surveillance: mainframes as well as desktop PCs, LANs, client-server software, telecommunication services, and also new products such as handheld computers within a corporate environment.

Focusing Information Technology and Telecommunication Investments

With ever-increasing pressure on IT management to deliver measurable results, not only the question of total costs has arrived but also the question of success and its measurability. Until now, there has still been discussion in the academic literature as to whether or not the use of IT/TC has an influence on corporate productivity and revenue. A very common thesis says that there is no relationship between investments in IT and company success. Nicholas G. Carr originally stated this, but several consulting companies have come to similar conclusions, when examining the annual IT investments as a proportion of the annual revenues in companies within the same industry. Besides this general discussion of the role of IT and TC as a whole, there also are similar discussions on single topics, such as e-mail, where productivity increases are denied or at least questioned.

Investments in IT/TC have common characteristics. It is very often simply not possible to prove success in monetary terms. Changing requirements and changing preconditions make this even more difficult to prove. The TCO view calculates not only the direct costs related to an investment over its whole life cycle, but also the indirect costs. Of course, not all cost components can be measured directly and many must be simply estimated. These estimates are difficult, since most cost components are hidden within larger general costs and are in no way transparent.

Up-front investments, such as buying hardware and software, as well as providing the necessary external and internal services, are in reality only a small part of the total costs occurring over the whole life cycle of the investment. The so-called unbudgeted costs arising over the life cycle, and not planned for up-front, usually make up the larger part of the total costs.

Cost Structures and Elements

Budgeted and unbudgeted costs must be differentiated. Direct (that is, budgeted) costs include the following:

- Communication costs (for example, costs for leased lines, external services, and WAN costs);
- Hardware and software costs, which include investments as well as leasing or rental fees for network devices, server systems and client computers, printers, and cabling;
- Management;
- Support;
- Development.

Indirect costs (or nonbudgeted costs) can be differentiated into two areas:

- *End-user costs.* One can find here all costs that occur at the end user due to time losses. This can happen if an end user experiences a problem and tries to help himself (without contacting support personnel) as well when he tries to learn things on his own or tries to implement macros, programs applications, and so forth, on his own.
- *Downtime-related costs.* This includes all productivity losses as a consequence of network or system downtime. This is usually measured in terms of the payment received by the end user over the time frame when she is not able to work.

The most prominent TCO concepts come from the Forrester Research and Gartner Group analysis companies. Since they differ in the details, they will lead to different results when used in the same scenario. Both companies provide a detailed list of all cost driving factors.

Gartner Group differentiates its factors into a number of different costs (budgeted and unbudgeted), which all belong to one of the groups [6] (see Table 5.4):

Table 5.4
Total Cost of Ownership

Capital	Administration	Technical Support	End-User Operations
Hardware	Asset management	Tier 1 help desk	Data management
Software	Security	Documentation	Applications development
IS allocated	Legal	Data extract	Formal learning
Network	Policy and procedures enforcement	Configuration review	Casual learning
Desktop	Formal audit	Application consulting	"Futz" factor
Server	Informal audit	Vendor liaison	Client-peer support
	Client purchasing	Standards development	Network costs
	Installation	End-user training	Network-peer support
	Capacity planning	Product introduction	Misdiagnosis
	Adds, moves, and changes	Product review	Peer training
	Upgrades	Newsletter	Supplies
	Server purchasing	User group	
	Security administration	IS desktop learning	
	Network operating administration	Planning	
		Utilization review	
		Install/move/upgrade	
		Service/PM (outsourced)	
		Install network	
		Tier 2 support	
		Tier 3 support	
		Technical training	
		IS network learning	
		Software distribution	
		Network operation systems maintenance	
		Disk management	
		Security/virus	
		Network operation systems configuration	
		Network operation systems performance management	

Source: [21].

- Capital costs;
- Administration costs;
- Costs for technical support;
- End-user operations costs.

The cost type listed as a "fudge factor" in Table 5.4 includes the abuse of corporate IT and telecommunications infrastructure for nonbusiness-related activities, for example, gaming and surfing the Internet.

Without knowing other TCO concepts, the Gartner Group list brings up certain questions that can hardly be answered. Most companies do not calculate their costs using this granulated approach. So all benchmarking between different enterprises based on this TCO seems to be somewhat questionable.

With a little bit of detailed knowledge, one can easily find missing cost driving factors in every published TCO calculation. If, for example, one examines typical VoIP TCO calculations, one will find that power supply and power consumption of the IP telephones are not listed as costs within most of them. The network (or the PABX/PBX) provides the power supply in traditional phone networks (with the exception of wireless devices), whereas every VoIP device needs a separate power supply. This can be provided per unit (possibly requiring additional wall power outlets) or can be provided as "power over Ethernet," which typically requires investment in new network switches supporting this feature. Although this is just an example, all total cost calculations carry a high risk of missing crucial elements.

Since the calculation schemes differ between Forrester, the Gartner Group, and several others on the market, the results will also differ, by a factor of up to 20% or more.

Other Cost Factors

Even when cost definitions are identical, there are other factors influencing costs. Within complex organizations, one can expect higher costs for planning, implementation, rollout, and operation than generally expected.

The type of usage also influences TCO calculations. User types can be differentiated into five types:

1. *Power user.* A power user is—from our perspective—someone who works with systems that are important for the company's success. These can be brokers as well as service technicians having direct customer contact or others. All power users largely depend on a working infrastructure and will incur high costs in downtime situations.

2. *Mobile user.* Employees in sales and field service are very often power users working with the more fragile mobile technology. The risk of technical problems with resulting downtime is relatively high, and the costs incurred by system failures and downtimes are also high.
3. *Knowledge worker.* A large group, but difficult to describe as an entity. All users collecting and processing information as a basis for decision making can be included here. The costs for downtime can clearly be seen but hardly calculated since they vary from task to task.
4. *Typical worker.* Within this large group are all users working repeatedly on the same or similar tasks, very often as a part of a workflow. Downtime costs are also variable, since typically only parts of their work need IT and TC support.
5. *Data typist.* This encompasses all simple data collection tasks. The downtime costs are relatively low, due to low wages and the fact that, depending on the type of failure, possibly no negative consequences occur, since, for example, a network shutdown does not necessarily affect data collection.

Given this list, one can easily see that unbudgeted costs for downtime are not only dependent on the services *not* provided, but also on the type of usage. Because most cost calculation efforts deny this fact, calculation errors must be expected.

User types, as just listed or similar, also help in explaining the differences between reliability requirements in different industries. For example, the retail industry very often does not invest much in the prevention of WAN failures. Because most systems within shops can work independently for hours or days, most retail chains can live very well with nonavailability incidents in their WANs. They usually find it sufficient to move sales or order data once or twice a day between the branch outlet and the company headquarters.

Within the supply chain of the automotive industry, the same situation could lead to disastrous results, since problems with "just-in-time" delivery can lead to massive problems in manufacturing and subsequent high costs. Also, nobody will accept downtime in online banking, brokerage services, or airline travel booking services. Investment risks are also neglected in common TCO concepts.

There are risks related to implementation, as well as operational risks that must be examined. The basics of implementation risks are simple: The newer and more complex the technology to be implemented, the longer will be the time frame necessary for implementation, and the higher the risk of failure. Implementation risks can be supplier related as well as user related.

Operational risks include these:

- Lack of performance (with or without downtime);
- Data loss;
- Security risks/theft risks;
- Jurisdictional risks.

Another important missing point in TCO calculations is "cost of change," that is, costs that are induced due to changes in requirements while a system or infrastructure is running.

TCO Summary

The target of every single TCO calculation is different and depends on the structure of the company. There is no single type of calculation methodology in sight. It is necessary to match TCO methods with corporate requirements, since there are different possible views of IT and TC costs. Knowledge of the common concepts is important for selecting the right approach.

Missing cost transparency within one's own company (especially when it comes to indirect costs) still limits the acceptance of TCO calculations within corporate decision-making processes.

5.4.3 Outsourcing and Cost Optimization

Decision making on corporate network infrastructure and service buying is very often dominated by a narrow-minded view. Most corporations only see the requirements of this particular need and try to fulfill this by buying from the service provider making the best offer in this particular case, but neglecting any expected future changes and extensions.

This behavior is typical when it comes to corporate WANs. A growth-oriented company will repeatedly select the currently cheapest vendor every time they need an additional data connection (leased line) or service. After only a few such repeated decisions, the consequences can be a loss of transparency caused by existing contracts with several vendors. The costs for every single component are as low as possible and even the direct costs are low, so this behavior seems to be good from a business perspective, but the whole picture changes when indirect costs are also calculated (such as cost of contract making and management, search costs for finding the provider responsible in case of a downtime, and so forth).

Of course, these strategies are not limited to WAN infrastructures. They can also be found in other segments of corporate IT and telecommunications investment.

5.4.4 Outsourcing and Business Value

> You can see computers everywhere but in productivity statistics.
> —Robert Solow, Nobel Prize winner for economics
> in *New York Review of Books,* July 12, 1987

This quote is also known as the productivity paradox of IT. It describes a major dilemma of IT investments: A business value very often cannot be seen. Business value is even more difficult to measure when it comes to the value of a network supplied by an external service provider and the network-related services. A simple cost comparison between costs for self-provisioning and costs for external services is much easier.

When it comes to infrastructure and related services in particular, value measurement is almost impossible. How can one measure the worth of the additional functionality a VoIP telephone provides over a traditional phone? How can one measure the worth of a faster reaction on an incoming e-mail due to instant mobile delivery?

In their book *Leveraging the New Infrastructure: How Market Leaders Capitalize on IT* [21], authors Weill and Broadbent describe four types of IT investment:

- Infrastructure related;
- Transaction oriented;
- Information oriented;
- Strategic.

According to them, the target of infrastructure-related investments should be the provisioning and securing of the infrastructure, as a basis for flexibility of the whole company or the company department. Besides flexibility, Weill and Broadbent see integration and standardization as the major value of infrastructure. One should focus on infrastructure cost reduction but should also bear in mind that these lay the foundation for IT services based thereon.

When it comes to transaction-oriented IT, one should focus on cycle speed and reduction of the costs occurring per transaction.

The major goal for information-oriented IT is increased control and information quality. Value can be derived from shorter development and marketing cycles, as well as better quality.

Finally, strategic IT can help increase market share and increase competitiveness, which will lead to more sales and higher revenue. While value creation can be clearly seen due to project risks, it is almost immeasurable up-front.

Although there are other concepts, such as total economic impact (TEI), presented in 1997 by the Giga Group analysis company [22], the basic dilemma of business value measurement of investments in infrastructure is not yet solved. The TEI model mentioned here focuses on evaluating the total economic impact of costs (direct as well as indirect costs, as discussed under TCO), usefulness, flexibility, and risk.

5.5 Service-Level Management and Service-Level Agreements

5.5.1 Service-Level Management

The provisioning of services by external service providers can lead to problems. On one side of the relationship is the customer who has requirements. His interests are driven by "time-to-market" considerations and he sometimes changes his mind. On the other side is the service provider—with limited financial and personal resources—who can deliver only in a limited way without compromising his business success.

The foundation for an unproblematic relationship therefore is always complete service-level management based on agreed minimum requirements, so called service-level agreements. Within traditional service-level management, only the availability of a network connection or server system is monitored. The minimum service availability should be at least as high as the percentage initially defined and laid down in the contract (for example, 99.9%).

This number-driven approach is insufficient from a user's point of view, because a service may be available but at the same time be too slow, have too long of a response time, and not actually be working in the way it should. Besides checking availability, other performance values must be examined. A typical suggestion would be to consider the following criteria, which are also called service-level objectives (SLOs) [23]:

- Response time at the user level;
- Response time of the server;
- Network-related delay.

One must also consider the type of measurement and the type of reporting. It remains to be discussed whether user-related measuring actually makes sense.

Services within the network can be examined using average measurements (for example, "average response time must be lower than...") or percent-related measurements (for example, "*X* percent of all transactions must have a shorter answer time than value *Y*").

Especially when there is only simple-average based reporting, single extreme values can lead to significant clipping effects. The actual numbers that should be used as borders (that is, what value the variable *Y* should then have in a concrete scenario) cannot be described in general, but depend on the user as well as the application.

When they are below a hidden limit, some delays are not even seen as such from a user's point of view. The limit is defined by user expectations. But there is still a limit—the amount of time where delays begin to "hurt." If the user feels this, she will stop working within the transaction or quit. The consequences of the delays, as well as the loss of productivity, are strongly negative. These limits are not only dependent on general user expectations and past experience, but also differ from application to application. While a user can accept Web page loading times of several seconds, the same user will not accept the same response time from an ERP application. To discover these hidden limits, one can perform tests and experiments but one should also directly ask the user groups.

By defining and permanently monitoring the values to be measured, and the limits to be examined, one moves from reactive to proactive service management. An important set of tools for service-level management is the Information Technology Infrastructure Library (ITIL) that also has best practices for infrastructure management. Most service providers already work on the basis of ITIL.

5.5.2 Service-Level Agreement

The criteria defined within the service-level management are written into service level agreements. The main goal is to describe the quality that must be delivered to the user but also to provide a basis for secure planning at the service provider side. Making things measurable and being able to measure them are necessary.

An SLA is not only a useful tool when buying external services, but can also be used internally between the IT department and special departments. The agreement itself is then much easier because it contains little legalese. In general, an SLA is a special agreement between a service provider and a customer. Within this agreement, service goals such as availability, answer times, reaction times, and times to rebuild) are defined. Costs for single services are also sometimes included in the SLAs. Negative incentives/fines if goals are not met are also very often defined. SLAs require a certain level of quality from a service provider but also protect service providers from user expectancies that are too high.

All of these detailed agreements are usually defined during the bidding process. Simple SLAs, which can often be found within the service descriptions of service providers, are very often not defined in a way that is helpful for the user. An SLA should include at least the following elements:

- *Service description.* Detailed description of the service provided.
- Agreement on *documentation* of all activities. This includes defining how things are to be documented.
- Definition of *responsibilities.* Is the customer, the service provider, or both responsible for a certain scenario or event?
- Definition of *interfaces* and connection points between customer and service provider.
- Definition of *goals.* Service parameters are defined here in detail An example for a typical description could be "Availability of network element *X* is 99.9%." All other service criteria (reaction time, time until all services are restored, and so forth) are defined here.
- *Service cost.* If service cost is not defined within the service contract, it is defined here within the SLA.
- Responsibilities of the *customer.*
- *Service times.* This can be, for example, MO-FR or MO-SA. Service hours are usually also defined 24/7 or 8 a.m.–8 p.m. To avoid potential problems, one should also include public holidays.
- Also included within an SLA is *reporting.* This will be described in detail in the following section.
- An *escalation scenario* is also usually defined in most SLAs. This clearly defines what can and will be done if the agreed-on goals are not reached. The consequences can range from review meetings to financial compensation.
- While *fines* (negative contractual incentives) can also be described within an escalation scenario, they should be explicitly mentioned separately. Contractual fines should be a last resort in any supplier–customer relationship and should compensate the user for the missing quality that the provider failed to provide.

The process of coming to an agreement on SLAs, apart from the not very useful standard SLAs suggested by service providers, often yields interesting insight into the world they live in and the way they work.

Very often, bargaining for the monetary specifications of contractual incentives and fines will lead to massive problems within the bidding process. The author knows of several requests for bidding from international companies, in 2004 and 2005, where the service expectations laid down were so high, and the fines for not meeting the goals were so high, that one must expect that the customer will attempt to use the fines as a way of paying for requested services, since he must know that the system requested will indeed have problems. These are also possibly the consequences of suggestions provided by some consulting houses that indeed suggest that the customer plays tricks to achieve unrealistically low bids.

5.5.3 Service-Level Reporting

Service level reporting is not a new discipline. In many cases, service-level management was seen as an "evil but necessary" part of the business. The historical reasons for this view were (1) mostly internal customers and (2) only a few criteria to be measured (such as CPU usage).

With the change in IT infrastructure to open systems, the conditions for reporting have also changed. Until now, service-level reporting has been regarded as necessary but still not liked, but the reasons for this are now different. Today, the sheer number and technical multitude of platforms are problematic because of the large amount of data that must be worked with. Current system management tools can be used to support this.

With an increasing trend to outsourcing, service-level reporting is receiving more attention as a tool for measuring the fulfillment of the services promised by the outsourcing partner. Even companies that did not previously measure service levels begin to do so when outsourcing services to an external provider, since they anticipate a possible loss of quality when services are provided externally.

Service-level reporting should consider the special contract terms of the outsourcing agreement. The following criteria should be measured to comply with what was defined in the SLAs:

- Availability;
- Performance;
- Stability;
- Answer times;
- System usage.

The report should be addressed to the needs of the groups targeted (customer, service provider, contract manager). The following common criteria are typically included: graphical view, monthly presentation, and provision in the language used within the customer's company (if this is a multinational company, then the English language is suggested).

Typical reporting problems include the following:

- Imprecise, incomplete, or simply wrong data (because, for instance, the wrong testing interval was used or problems were experienced with report generation and synchronization between different platforms);

- Excessive use of average values instead of minima/maxima and other criteria;
- High workload for generating reports.

A monthly report can contain the following elements:

- Addressee;
- Communication systems;
- LAN graphical performance view;
- WAN graphical performance view;
- Security incidents;
- Workload profile;
- Trends in resource usage;
- Help desk services;
- Help desk call logging;
- Help desk call response;
- Help desk problem resolution;
- Help desk statistics;
- Mainframe services;
- System support overview;
- Mainframe availability;
- Mainframe performance;
- DBMS performance;
- Storage system performance;
- Print job performance;
- CPU usage;
- Midrange services;
- Application availability;
- Summary.

This list merely provides an idea of what should be included. The concrete realization varies from company to company.

5.5.4 Incentives and Fines Within Outsourcing Relationships

As with other "enabling" technologies, network consolidation has a different relevance and gravity from company to company. A major precondition is the understanding of one's own processes and requirements before investing in the integration of voice and data networks or using a supplier service. The following questions help to determine the importance of network consolidation and communications resourcing for one's own company:

How many company locations are there and how many of them have users with technical expertise (IT and TC)?

Which services/technologies would help to make the working life of your users easier?

Which functionalities of a convergent environment are the most important for you (unified messaging, videoconferencing, Web-based telephony, always-on, and so forth)?

How large is the share of your business that is driven or influenced by high-volume communications with your customer?

How does your customer rate these communications and your customer service?

What would you do to make this better (provided that there are no technological limits)?

How adaptive to change/how scalable is your current IT data network/ PABX/PBX infrastructure?

How can your infrastructure be protected from external influences?

How mobile are your employees?

What proportion of employees are working from home?

Which communication channels have been the most important up to now?

Providing answers to these questions helps to clarify the importance of individual criteria in an actual outsourcing scenario.

Positive and negative incentives have different influences. While negative incentives and fines should stop unwanted habits, positive incentives should be paid when goals are achieved. Management people know and use these ideas to motivate and evaluate the performance of individuals or working groups. If basic expectations are not met, there will be negative consequences. If expectations and goals are met or overachieved, however, bonus programs, career perspectives, and sometimes company share bonus programs should act as a thank-you.

Similar principles work in supplier–customer relationships when buying services. If goals are not met, the negative incentives must function immediately and hurt enough to provide an adequate deterrent. On the other hand, the incentives for achieving goals must be high enough to encourage the provider to invest the necessary financial and personnel resources. Unfortunately, most public discussion on the topic only relates to negative incentives and fines.

Negative Incentives and Fines

Negative incentives and fines are very often misunderstood in outsourcing relationships. The first goal should not be to punish the supplier or to compensate for the financial losses incurred by nondelivery or poor delivery of services by the provider. Instead of this, agreement should be made to prevent problems from occurring and—if this basic goal cannot be achieved—help problem solving for the current situation and improve things in the future.

But how can one discover the right level of negative incentives that should be written into the contract? The minimum amount should be at least higher than the administrative costs related to paying and receiving the fine. One should also define an increase in the penalties for each time the same problem arises again, up to a defined maximum amount. This maximum amount is typically limited to a percentage of the monthly bill. If the goals are repeatedly not reached, the final sanction should be cancellation of the contract. All payments should be made immediately, or as soon as possible, so that there is a timely relationship between not reaching a goal and the penalty for this. The fines should be paid on at least a monthly basis.

A number of different criteria can be used to determine negative incentives and this can be done in different ways. Single services within an outsourcing scenario are treated in order of importance. Fines can now be determined as a function of importance, relative to the monthly bills. In some cases, fines are related to trends. In this case, a service that continually fails will be sanctioned and *not* a single negative event, since everything else runs fine. Other fines can be related to, for example, provisioning of single employees (the ones with important knowledge!) but are not very common when it comes to network outsourcing. After all, it is important that, after all deliberations and planning, only business-relevant incidents and criteria are supported by negative incentives. If, for example, access to a centralized ERP system is not possible for a few hours due to a router failure over the weekend, this would lead to serious lowering of the availability percentage rate when viewed 24/7. From a business point of view, the incident is probably not relevant, since nobody would have worked during the downtime anyway and only the monitoring system knew that there was a problem. This simple example shows that a fixation on availability figures alone can lead to problems.

Another aspect of relating sanctions to downtime can be found within the definition of downtime itself. Sometimes, up to 60 minutes per month of downtime are allowed. So the fine must usually be paid from minute 61 onward. Of course, this should not be a flat fine, so 61 and 60 minutes, which are quite similar, would be treated extremely differently. Hence, a distinction needs to be made between one 60-minute downtime and 20 downtimes of 3 minutes each. These differences are very often not described in the agreement but the consequences for business can differ enormously (depending on the company business model).

The solution can lie in a varying scale of penalties. The consequences can be related to, say, the number of users, the number of locations with problems, the duration of downtime (scaled as follows: up to 1 hour, 2 to 4 hours, more than 4 hours, and so forth), or frequency (first, second, third incident during this month or quarter) and measured with point values. Most first-time users would find this amount of work somewhat suspect, but one should take the same care in defining sanctions as in defining the services. The same care should also be taken when it comes to monitoring by the customer and the positive incentives described in the next section.

Positive Incentives

Positive incentives should be a thank-you for achieving goals and positive habits, where payment is made for service quality increases that improve business for the customer company.

Generally speaking, the service provider should receive incentives for identified cost savings, service quality increases, and other activities that increase the customer's success. Positive incentives are useless when linked only to the contractually defined services. Why should the customer pay extra for that which is already paid for within the service contract? Simple overreaching of defined goals normally generates no additional value.

Bonus payments, therefore, should be linked to the achievement of specifically defined targets. These can include delivery within a certain time frame before the defined date or in the increases of business-critical values. Financial compensation should only be paid when there is an increase in business value on the customer's side.

Also included within a positive incentive program can be a positive development in (measured) customer satisfaction or—when buying services from multiple suppliers—better performance of one supplier compared to others delivering the same or comparable services to the corporation.

Very common in traditional outsourcing scenarios are agreements defining sharing (financial measurable) effects, for example, due to a decreased error rate in production in a defined way (for example, split 75%/25%) between customer and service provider. This type of sharing will probably not be achievable in network outsourcing due to the difficulties inherent to measuring the financial consequences.

Consequences

Positive and negative incentives can be an important way to help establish and continue a supplier–customer relationship when outsourcing a network infrastructure and related services. Lasting success is only secured by continuous and consistent use of a system of incentives and penalties. Nothing erodes faster than the search for excellence, when the supporting systems are not adhered to.

5.6 Selecting the Right Service Provider

Once the decision to outsource has been made, the vendor selection process begins. The workload for planning and implementing vendor selection will vary, depending on whether the entire voice and data infrastructure, or only a part of it, is touched by the decision.

The path to take is different from company to company. The following description includes typical paths to take and tries to provide hints that are not only important for the success of the bidding process itself, but are also important for a successful long-term relationship with the service provider. To achieve this, it is most important to steer the process right from the beginning and to adhere to the line established from the start without moving too much in a direction a vendor might suggest.

5.6.1 The Right Supplier

The following types of supplier are potentially relevant when outsourcing network infrastructures and infrastructure related applications:

- Service department of a hardware supplier;
- System integrator/specialized outsourcing service provider;
- Traditional network service provider/carrier with outsourcing offering;
- Technology-oriented business consulting company;
- Specialized ASP/hosting service provider;
- Value-added reseller or team of different vendors.

Depending on what matters most within the planned activities, the selection can be limited to special groups or single vendors within these groups. Companies described within the case studies in this book (see Chapter 6) typically say that they talked with no more than six to eight vendors, since no more than these were seen as suitable.

5.6.2 Considerations When Selecting a Service Provider

To assist with a systematic preselection of the vendors available on the market, several criteria are listed here. These criteria must be fulfilled to meet the basic needs and requirements. The following 15-point list puts together these criteria to aid in a simplified preselection process:

1. Searches should begin with current suppliers. These providers know the customer well and will try to remain in business.
2. One must be careful when comparing the different offers from different providers with one's own requirements. Several suppliers offer only limited services or service qualities (for example, low-cost VPNs over the public Internet) and try to push their offers without paying much attention to the needs of the customer. A provider with a larger portfolio and variable service offerings and levels for different locations in the WAN will probably be the better partner in the bidding process.
3. Network coverage. The network footprint of the provider's own network should not only cover all current locations but also all those that are potentially relevant in the future (potential markets, potential production plants).
4. Checking of regional infrastructures. The more locations the provider has near to the customer locations the better, since connections will probably be less costly.
5. Checking service offers for teleworking. Vendors with their own network structures have advantages here.
6. Checking technical information about the provider network. One should also look at the ownership of the network or parts of it.
7. Checking compatibility with one's own network equipment (as long as it is to be used within the new contract; if it is going to be replaced, then this does not matter). Small providers support a limited number of different devices.
8. Finding information on vendor processes and the way the company works.
9. Taking human nature into account. The "human infrastructure" makes the difference. Who is the one sitting opposite you in the bidding process? What answers are given when during the process you ask

the personnel working on the agreement? Do the personnel speak the same language or at least English adequately? Is local support available? In the local language? Does the vendor have a "one face to the customer" approach?

10. Security risks must also be taken into account. Has the vendor recently had any security problems that were noticed in public? Is the vendor's data center and network operating center (NOC) secured or can one simply walk in? What security concepts are provided by the vendor?
11. Are there any network monitoring tools and what information is included with them? A service provider should give you a toolkit that allows you to monitor the status of your network, applications, and devices at any time by yourself.
12. Ask for proactive monitoring. The vendor should not only try to restore a service or a connection after receiving a customer complaint, but should also solve any problem that arises without the customer needing to become involved.
13. Checking SLAs. A detailed examination of the SLAs suggested by the provider shows how strong the provider acts in his service-oriented thinking. One must consider every single word in the SLA description. It is best to develop your own descriptions derived from your own business requirements and compare them to the SLAs offered by the vendor.
14. Checking references. To validate references given by the provider one should always communicate with the contacts mentioned at the company that gave the testimonials. One should not believe a single word without doing this.
15. Checking the financial preconditions. Several vendors, especially in the area of telecommunications and networking, have suffered severe financial shortages, gone into bankruptcy/Chapter 11 in the past, or are on their way to doing so.

Depending on the requirements, these aspects must be considered and evaluated. With the help of a simple scoring process, one can find the potentially best providers. The selection process should then be started in detail only with those providers who pass this initial preselection process.

5.6.3 Selection Process

The provision of external services, especially when it comes to things such as managed services, should never be seen as good or bad but rather from a viewpoint that is as neutral as possible. Positive opinions and perceptions are very often vendor driven or come from other participants such as contract lawyers or consultants. In the best cases, these opinions come from companies who have already taken this step and are satisfied with the results. Negative comments are often received from companies who are not satisfied with the results, from organizations and their employees who will be directly negatively affected, and from employee and union representatives.

A project of this kind should only be started when at least one of the following expectations exists: (1) better service quality with the same cost or (2) same service quality with less cost. The TCO mechanisms already described in this book will help you find the right approach. The most important deliberations are listed in the following subsection on planning. In the subsection on the negotiating process, a customer-centered concept for steering the bidding process between customer and service provider is presented. The aspects that are potentially relevant to a future vendor change or a return to self-provisioning and operating are also discussed.

Planning

Planning for such a far-reaching outsourcing decision should integrate business aspects as well as legal aspects. The best foundation for success initially requires a detailed internal analysis of at least the following areas:

- A precise description of all services to be included that is derived from what is presently done by the company itself.
- Parallel to this, a preselection of all potentially relevant vendors by using secondary information (market studies, analyst materials, consultants).
- Listing of all personnel and technical resources that are already used for providing services (within the old scenario). Most companies already discover here that they still have no complete overview.
- Definition of service levels and balancing them with historical data to discover how far these met in the past (this implies a great deal of work if the data was not gathered in the past).
- Calculating the total costs of providing services on one's own (if the activity was previously outsourced to another vendor then this must also be calculated). The approach should be as complete as possible. The TCO chapter in this book can help to define the individual criteria.
- Define the minimum expectations of the service quality required from the external provider.

The results of this internal study now provide what a vendor would call a *baseline.* If a vendor can deliver the same results at lower cost, or with better results at the same cost (or both: cheaper and better) than the company itself, then the answer is *yes,* external service providing makes sense from a business point of view.

With this internal baseline, the company now has the necessary foundation to issue a request for proposal (RFP) to the vendors that were previously preselected. This RFP should not only include a detailed description of the internal analysis and pave the way for contract negotiations but should also define the goals, or at least the general conditions, for a provider offer. The RFP includes the following:

- Technical description and details;
- Financial expectations;
- Preconditions of the negotiation process;
- References to a letter of intent (if this tool was used).

Without an in-depth internal analysis as a foundation, it does not make sense to provide an RFP. When problems occur within an outsourcing relationship that lead to a higher workload/higher costs, and consequently to dissatisfied customers, very often the problem can be traced back to an incomplete or error-prone analysis phase. It can be useful to rely on an external consultant here, if the necessary experience is not available within the company or cannot be made available.

The RFP should therefore include a precise description of all resources used, as well as the services that must be delivered: all elements of the infrastructure, network architecture, hardware, software, operational systems to be used, and all personnel resources required as well as a functional description and a definition of responsibilities.

The list should be detailed enough so that a potential supplier does not need to come back and ask about details or later discover that he needs to change (typically increase) his offer because he initially had insufficient or incomplete data that caused him to offer the wrong things.

Of course, these recommendations are easier to write down in a book than to actually implement. However, when owning the process during the whole bidding and negotiation phase with the potential suppliers, it is not only useful to follow these recommendations, it is essential.

Criteria for Selecting Service Providers

For the selection itself, scoring models are typically recommended that validate selected aspects with different amounts of points and lead to a sum of points for each vendor. Depending on the path taken, not only hard but also soft factors must be considered. These factors can include the following:

- Fulfillment of required services;
- Price/costs (TCO oriented);
- Services (what is offered);
- Provider footprint;
- Proximity of provider locations and own locations;
- Trustworthiness;
- Financial solidity;
- Project management competency;
- Partnering/certifications with and from all relevant players;
- Vendor know-how (of activities that need to be outsourced as well as general industry knowledge and experiences);
- References, testimonials, customer satisfaction (all testimonials should be checked);
- Flexibility, relationship quality, accessibility (a single point of contact would be ideal);
- Possibility of customizing the services offered;
- Data protection/data security;
- Backup concept;
- Disaster recovery planning;
- Existing SLAs;
- End-to-end service guarantees;
- Other risks;
- Inclusion of third-party services.

The importance of the aspects to be considered when selecting vendors not only differs from company to company, but also from outsourcing scenario to outsourcing scenario. Hidden behind this process is always the relationship between cost and risks, which are different depending on their importance to the success of the company.

Negotiation Process

Gaining control over the process is the major goal of the negotiation process planning. In-depth preparation that leads to a clear definition and communication of a way to go, starting with an RFP, is the major success factor for an efficient negotiation process and for the minimization of the imminent risks of contracting.

Legal aspects also touched on here are described elsewhere. The methodology described here also includes suggestions presented by Gordond & Glickson. These, and other consultants, now suggest preparing not only an RFP but also the contract itself on the company's own and to not wait for a contract presented by a supplier.

Of course, there is a lot of work to be done and one will probably need help from a lawyer with a focus on contract law in doing so (since many companies do not have their own legal department). This contract should be presented to the vendors before the negotiation process is started. While this generates quite a workload, it is the best way to steer the process in the intended direction. One should not stick to the plan, however, without considering changes. But any vendor who wants to change parts of the contract is requested to provide a structured change approach and to document any wanted changes.

When it comes to the negotiations themselves, a clear view regarding the process time can be helpful. In an ideal scenario, one would limit the negotiation days and even the hours per day when negotiations will occur. This helps both parties reach the goal of success. It is very important never to leave the track once it is started.

When it comes to the description of the service itself, very often legal help is needed because the selection of single words can influence the way potential loopholes can come up or be closed and possibly later be used as a vehicle for the provider to charge for additional costs.

For those who have never written a service description before, they will have a hard job doing so (for both the RFP and the contract itself). It is helpful to demand a service description as the response to the RFP from the service provider. During the negotiation process, this service description should be examined in detail and changed if necessary. The contract itself, or general agreement, should include all relevant aspects of the relationship. In addition to this, the service description pricing agreements and price lists must also be included. When negotiating a contract that is intended to last for several years, one must consider that pricing will change, at least for parts of the service delivered. The prices of networking and related services usually decline year after year. So, from the point of view of a several-year contract, prices will go down during the contract period since costs for hardware (for example, for routers, VoIP devices), network connections, and so forth, will decrease. A "best-price" agreement is very often included but this does not always work in the intended manner because the basis for relating the best price is often not defined or remains unclear. Not every vendor will relate its pricing to a "market price," since this price will be endangered by special offers and price dumping by other vendors. While no general recommendation can be given here, one should nevertheless keep an eye on the definition of a reference price or reference price level.

Price reductions are not always the most important issue. When speaking with several companies who have already outsourced their network infrastructure or parts thereof, or who are currently negotiating this, one can find another important driving influence: Some companies mention that, instead of looking for cheaper lines, they are more interested in getting more and more connectivity within a constant budget. This is good news for the providers and secures their revenues.

When reading this book as a representative or sales force member of a service provider, one might want to put a curse on the author and his recommendations, since at first sight he recommends hard negotiations and price discussions. However, this opinion does not do him justice. While several consultants suggest, even in the public, making the RFP information look better than it really is to get better conditions, the author does not view this as a viable strategy for achieving a satisfactory long-term relationship.

Instead of trying to cheat, one should find a way to reach a fair agreement, since experience shows that if one partner feels himself to be the loser or to have been wrongly informed on important aspects, he will use the next possible situation as a basis to charge more to compensate for the losses he saw in the original intent.

Sticking to the original steps listed earlier also helps when negotiating with several vendors at the same time, but can sometimes lead to the scenario that only one player remains while others, not willing to cope with the strict requirements, will give up.

Whether or not there are one or more potential suppliers involved, it is necessary to define a process for limiting requested changes in the content of the agreement. Every change requested by a potential supplier must be outlined as such and needs qualification. This makes the whole process easier, since any changes can be tracked back. If a vendor must explain why he demands a change, this will also lead to fewer change requests in total. When negotiating with several vendors, this also helps to keep track of the process, since no vendor wants to be listed out due to changes demanded by him in single terms of the contract.

The negotiation process itself should be done according to a clear timetable. The number of days for negotiation, the number of negotiation rounds, the number of negotiation hours per day should be decided ahead of time. Every vendor should be treated in an equal manner. No exceptions are allowed here. If this is initially clearly communicated to all potential suppliers, then the approach will work, and no supplier will try to break out of the process and change the rules of the game.

Adhering to this process can mean conducting round 1 with vendor 1, then round 1 with vendor 2 and round 1 with vendor 3; then round 2 with vendor 1, round 2 with vendor 2, and so forth. It is best if only one team speaks with all vendors. This will save time, since communication between teams is not necessary and, as far as its possible, it ensures equal treatment of all vendors.

This type of customer negotiating team needs different skills (technical, financial, jurisdictional) and therefore must be staffed with team members from the following departments:

- IT/TC management;
- Technical services;
- Finance/sourcing;
- Legal department;
- Human resources (if employees are to be moved to the vendor).

In an ideal situation, all bidders should provide their change requests in round 1 (which of course makes it necessary to provide the basic contract with the RFP, as recommended earlier). Considering further change requests can also be useful before beginning round 2 of the negotiations. Results should be available at the end of each round and, by the end of the last round, every change request should have been discussed and a solution found. After all rounds are finished, the providers should be given some time to provide the final offer. This is very important for final pricing of offers. All changes that are accepted should be written down in a document that must be given to the provider and which will later be part of the agreement. Adhering to this type of rigidly defined process will hinder the vendors from attempting to lengthen the process and gain control over the way it is done.

The workload implied by this process means that most companies will be reluctant to provide their own contract and will try to use the agreements supplied by one particular vendor, but there are risks in doing so. This type of vendor-delivered contract will be more or less vendor oriented, down to the last term and detail. This is not a bad thing, rather a normal part of business life and is acceptable as long as the vendor accepts negotiation of changes to the important parts. Problems arise mostly for small and medium enterprises, which often accept such a contract in the form already provided and focus their negotiations only on the price and other side aspects.

Planning for a Possible Termination

When the end of a contract is near, whether it was written by the vendor or the customer is of little importance. Simply "letting it run out" will not work since the infrastructure involved touches central aspects of the company's ability to function. In the case where a prolonged relationship is no longer possible or no longer desired, a mechanism must be defined to allow an exit from the contract and migration to another service provider or a return to self-provisioning as was done before the provider was selected.

One must always allow for a potential bankruptcy/Chapter 11 action by the service provider. It is recommended that you secure the most important assets within the contract and take care that, in the case of the financial collapse of a provider, the services can be continued in a useful way and that you avoid the risk of entering into an unwanted relationship with other companies (who may buy themselves into the contract after bankruptcy of the original provider).

5.6.4 Contracts

The most important part of any business relationship is the contract. Similar to a marriage contract, one begins with the clear idea that one will never have to use it against each other. But, just in case, one must be careful from the outset. Standardized contracts are not particularly useful because all relationships have individual components and a standardized contract is usually a too general to cover all situations that might arise.

The following aspects should be included in all outsourcing agreements: (1) what is included, (2) cost, and (3) service level provided. The contract should be customized as required and must cover all of the possibilities that can arise over the duration of the contract.

A common understanding of what is to be delivered and what is to be done if problems arise is extremely important for the success of all outsourcing activities.

Negotiations and Letter of Intent

When it comes to detailed negotiations, a letter of intent (LOI) can help define the area for finding an agreement. An LOI is an agreement written down before the contract is negotiated and secures the relationship before signing the final contract. This can be useful because, when outsourcing, an external service provide will probably need deep insight into the business internals of the customer in order to calculate the project requirements.

Both sides experience costs during the negotiation process. A possible billing or partial billing of services delivered before the contract should be clearly included or excluded in the negotiations.

Contents of an LOI

An LOI can include the following:

- Definition of general project goals with contract details in mind;
- Expected outcome of the negotiation period;
- Range and criteria for cancellation;
- Exclusiveness;
- Power of attorney;
- Definition of how far the LOI is binding;
- Up-front definition of what is provided by whom and whether or not compensation will occur;
- Confidentiality agreement;
- Agreement about not trying to acquire customer employees;
- Responsibilities;
- Definition of who holds the rights to any results;
- General terms and conditions.

Services Delivered Before the Contract Is Signed

A question that needs to be addressed relates to the services that must be provided before the contract is signed. These can include the following:

- Detailed studies;
- Consulting and consulting duties;
- Definition of a migration concept;
- Definition of the time frames and milestones;
- Integration of external services;
- Duties of the company requesting the bid.

Related to this list is the question of how services will be paid for and whether the rights to the project outcome are available, especially if the negotiation process breaks down before a contract is signed.

Depending on the content of the LOI, this can be defined quite differently. One should therefore clearly define any binding terms, as well as any payments, within the LOI.

Negotiation and Fixing a General Agreement

It is recommended that you use a general agreement as the foundation of all contractual relationships with the service provider. This general agreement defines the services to be delivered, including all possible components, even if they are only ordered or canceled later during the project. Only with this concept can the necessary flexibility be included during the 2- to 5-year period of the main contract. With such a general agreement, individual tasks do not need to be redefined when they are needed.

It is also possible to include a best price term within this agreement to make use of future price decreases. Increasing prices are not to be expected anywhere in an infrastructure and related services anytime in the near future.

This type of master agreement can include the following terms [24]:

- Contractual basics;
- Service definitions;
- Securing the service;
- Rules for living the contract;
- General terms and conditions.

The contractual basics include these:

- Definition of the goals of the relationship;
- Contract parties;
- Transferability, allowance of subproviders;
- Planning and system responsibilities;
- Requirements for testing;
- Technical, organizational, and financial framework for the contract;
- Quality criteria;
- Definitions of terms used.

The service definitions include these:

- Phase planning for migration and transfer of service to the service provider;
- Takeover of employees;
- Takeover of running contracts with third-party suppliers;
- Achievement of service levels;
- Help desk and user support;
- Availability;
- Reporting;
- Documentation;

- Installations;
- Data transfer and access rights;
- Maintenance;
- Trouble detection, trouble ticketing;
- License rights;
- Payment duties;
- Duties of the buyer.

Securing the service includes these tasks:

- Quality assurance and monitoring;
- Customer satisfaction index;
- Warranties and responsibilities;
- Incentives (positive/negative) and fines;
- Guarantor relationships;
- Protection rights;
- Confidentiality agreements.

Rules for living with the contract include these:

- Project management;
- Working groups;
- Change requests;
- Acceptance of services;
- Rights regarding cancellation;
- Contract end and insourcing;
- Handbook on how to work;
- Duty to cooperate.

How to Define a Service-Level Agreement

From a legal view, at least the following components should be included in a SLA:

- Technical as well as content definition and definition of sums;
- Duties of the buyer;

- Interfaces;
- Definition of service quality;
- Reporting and monitoring;
- Rules on payment;
- Negative incentives and fines;
- Duration of the SLA agreement.

SLAs are useless without a precise definition of system parameters with which the system performance can be measured. Before talking in detail about reaction times and availabilities and giving numbers (for example, 99.8% availability, maximum 4 hour of downtime until everything is working again), one should first take a closer look at the definition of the terms. The following list describes the most important service parameters:

- *Operation time/system operation time:* the time frame over which the systems and supporting services are available. One must deduct maintenance time and time frames reserved for necessary changes and upgrades.
- *Service hours:* time during which the service is proactively monitored and support calls are answered. Within the service hours, incident and problem management works to resolve any problems that occur.
- *Maintenance time:* The provider needs the chance to work on the IT system (for example, for changing releases, changing configurations). Because maintenance times are excluded from the operational times, planned maintenance does not compromise SLAs.
- *Reaction time:* the time frame between detecting a problem and reacting to it by trying to diagnose it. The actual reaction time is measured on the basis of the problem management tool. Reaction times are only measured during service hours.
- *Time until system is available again:* time between detecting an incident and when repair is finished. This is measured based on the time detection of the problem management tool, even if the person who originally detected and reported the problem cannot be reached with this message.
- *Availability:* measured based on the yearly operating time minus the time of nonavailability. This information is given as a percent.
- *System availability:* describes the availability of the system as a whole. Downtime of a component that is built redundantly will not compromise system availability.

The parameters listed here are examples taken from daily work, but the definitions can differ from vendor to vendor.

If one measures the provider quality with criteria that go beyond simple technical measurements, for example, by asking end users about customer satisfaction, it is recommended this be placed in a separate or additional agreement.

Exit: Insourcing If the Project Fails or When the Contract Expires

Contractual rules are not always sufficient to protect a project from major failure. Depending on the project, one should plan early for a possible exit by completing these tasks:

- Securing access to all data;
- Making claims for the return of data and program files;
- Securing all software licenses necessary for doing business;
- Ensuring the existence of a service delivery manual;
- Obtaining access rights at the data center;
- Obtaining the right to take over hardware (such as network components);
- Determining the duty of cooperation with third parties (for example, the new service provider);
- Determining the duty to continue support;
- Planning for emergencies.

These exit-oriented agreement terms also help when the relationship goes well but the service provider goes bankrupt. Returning to self-service is also known as *backsourcing*.

Other Parts of the Agreement

Change Requests

Change requests are very often problematic. To avoid lengthy discussions regarding changes, one should deal with the potential problems at the outset:

- Define who is allowed to request changes.
- Describe the consequences on other services, potential consequences on times, the duties to assist the change, and also the terms of payment.
- Confirm that change requests can only be provided in writing (this is important if problems occur later on), and that a request will only be followed once both parties have agreed to it.

Inventory Management

An inventory management system without gaps is important for outsourcing of network infrastructure. The necessary terms and conditions should always be included. Because network outsourcing always (with the exception of greenfield installations) starts with a takeover and changes in existing infrastructures, it needs to be clear from the outset who is responsible for a primary inventory check and who assumes the costs.

Within an inventory management system, the inclusion of other suppliers is possible. Of course, interfaces must be defined to ensure permanent, frictionless operation.

Target Quantities

Target quantities define a target corridor within which the customer may place an order. Target quantities are very important as a calculation base for the supplier. These target quantities are derived from the original planning and any changes (increases and decreases) agreed on thereafter, and they are normally used over the entire duration of the contract. The target corridor is typically defined around a target quantity that may vary by +10% to –10%. Within this target corridor, the customer can order services without compromising the underlying pricing scheme.

Obligation to Cooperate

The duties of the customer must also be written into the agreement. This is not solely in the interest of the supplier but also for the customer. The most important duties are (1) supplying the necessary space and power for all equipment provided by the supplier that must be placed in customer locations (network components, server systems, and so forth) and (2) provision of access to these spaces during regular service hours (can also be 24/7).

Order Processing

The agreement should also define who (which person) on the customer's side is allowed to order services from the provider. The way in which orders are made should also be defined. This is most important for correct billing afterward. Online ordering within the intranet of the user (or a provider extranet or service portal) is an additional way—or sometimes the only way—of placing orders, and sometimes (if the necessary functionality is supported by the supplier) also a way of directly making changes such as selecting ports or changing the bandwidth of an existing connection.

Migration Planning

Because most companies suffer from repeated migration of individuals, workgroups, or entire teams throughout one or more company locations, a certain number of changes should be included in every contract, as well as a standardized way of ordering migration support from the supplier.

Common Criteria

One should always establish a path for growth within an agreement, not only with migration in mind, but also with a view to the future. The infrastructure provided by the supplier should be at least one size larger than currently required, to allow for additional users and usage. Rules within the contract usually define how many ports are reserved for additions, as well as how much extra bandwidth than necessary is provided or can be immediately provided. These target values are typically described per network segment and device, but can also be described as a general agreement that affects the entire infrastructure.

References

[1] Hagel, J., and M. Singer, "Unbundling the Corporation," *Harvard Business Review*, March/April 1999.

[2] Harbison, J., and P. Pekar, "Smart Alliances," San Francisco, CA: Jossey Bass Publishers, 1998.

[3] Prahalad, and Hamel, "The Core Competence of the Corporation," *Harvard Business Review*, May/June 1990.

[4] Pierre Audoin Consult, 2004.

[5] "Update on the Outsourcing Market: Trends and Key Findings," Dataquest, 2004.

[6] Gartner Group, 2004.

[7] Diamond Cluster, 2002.

[8] "Explaining Capital Market Success of Financial Service Outsourcing," E-Finance Labs, Frankfurt University and Darmstadt Technical University, 2004.

[9] "Outsourcing of IT—An Option for SMEs," Munich: EFOPlan Group, LMU University, July 2003.

[10] Riedl, *Information Management & Consulting*, Vol. 8, 2003.

[11] St. Gallen University, Center for Outsourcing Studies, Switzerland, 2004.

[12] Deloitte Research, December 2003.

[13] "Offshore Outsourcing," Datamonitor, June 2002.

[14] Gordon & Glickson, www.ggtech.com, 2003.

[15] Global Insight, April 2004.

[16] *Wired* magazine, July 2004.

[17] "European Business Process Outsourcing (BPO) Forecast and Analysis 2001–2006," IDC.

[18] "Taxonomy of Internal Business Processes," Yankee Group, 2003.

[19] Compass Group, "Global vs. Regional Telecommunications Sourcing," *CIO Magazine*, July 29, 2000.

[20] "Outsourcing Your Network Managed Services—Key Considerations for Success," Infotech, 2002.

[21] Weill, and Broadbent, *Leveraging the New Infrastructure: How Market Leaders Capitalize on IT*, Boston, MA: Harvard Business School Press, 1999.

[22] Giga Group, 1997.

[23] "Optimization of Network Applications," *Funkschau*, August 2004, pp. 48–50.

[24] Presentation of Heussen Law Firm, October 2003.

6

Communications Resourcing

The future is here. It's just not widely distributed yet.
—*William Gibbson*

To cope with increased complexity, companies are looking for external help with provisioning and operation of IT and TC. The arrival of so-called "disruptive technologies" such as VoIP and MPLS VPNs are always a reason to examine whether outsourcing is an option. Communications resourcing provides a new way of doing this.

6.1 Concept

"Doing more with less"—one can sum up the pressure on corporate IT/TC with this simple sentence. This is especially true for the infrastructure of voice and data networks because it is difficult to measure to what degree they are relevant for corporate success.

The technological changes that brought us the convergence of voice and data networks also provide a radical new concept for purchasing voice and data networks and related services: *communications resourcing* (CR). CR helps us cope with concurring and complex demands on corporate infrastructure. It provides costs and utility estimates but also risk minimization by avoiding the weak spots of usual outsourcing strategies.

The core factors of communications resourcing are as follows:

- Communications resourcing sees voice and data networks not as a core competency of one's own company, and demands the cessation of all self-provided voice and data networks as well as the related hardware, software, and services. What once was self-provisioned and operated is to be given to an external service provider on a step-by-step basis.

- Communications resourcing is based on simplified provisioning of voice and data services on a convergent network. CR demands and needs a unified infrastructure for voice and data.
- Communications resourcing needs standardized service-level definitions and standardized support that is suitable for the whole infrastructure.
- Communications resourcing includes permanent monitoring and proactive management of the entire infrastructure.
- Communications resourcing requires variable and transparent cost structures.
- Communications resourcing requires unified billing (that is, one bill for all services).
- Communications resourcing relies on single sourcing.

Communications resourcing requires a company to give up self-provisioning of voice and data networks due to the fact that a highly specialized service provider is able to fulfill those network needs, including security management, better and at lower cost due to economies of scale. When it comes to cost, the most important factor determining cost savings is the unification of the infrastructure. (Potential cost savings and additional benefits were discussed earlier.)

Communications resourcing also requires unified SLAs and centralized, clearly defined service and support. This requirement addresses another important weak spot of corporate networking. The faster a system or network is up and running again after an incident, the less business processes are affected. The SLA needs to clearly define what level of quality is expected and must be ensured. This should be a single definition for the entire network but of course it is also true that some parts of the infrastructure are more important than others. For example, the main LAN connection at the company headquarters is much more important than a connection to a single teleworker or a small branch office. So, from an economic point of view, one would normally come to the conclusion that a prioritization of services and support is necessary while availability should be at the same high level throughout the entire network. To solve this dilemma, communications resourcing usually includes a master SLA agreement that not only addresses the different needs in different parts of the infrastructure, but also includes a unified support structure.

Network management should be done proactively. This means that the provider should detect and address any network problems immediately, without the need for the customer to report the problem. The goal here is to detect the problem and solve it before the customer even knows that a problem ever existed. Only with proactive management can the quality necessary for an externally operated network be delivered in the way that it should be—at least in the long run. Although the network is completely managed by the provider, all information on the status of the network should also be accessible to the customer and the best solution for this is a Web-based customer service portal. Although there is no technical need to do so, a customer service portal can help relieve the worries a customer has when surrendering control over the network.

Communications resourcing also provides more flexibility. The costs are largely related to operation only. If one follows the CR concept, changes or relocations are not a problem. Changes do not lead to fixed costs, as was true from early networking when adding an additional connection led to the necessity of buying a new switch when all connections the old one could handle were already used up. Also, high change costs associated with, for example, PABX systems, in which anything that differed from the original setup led to high service bills, no longer exist.

A single unified bill, with a detailed and precise listing of all service components, should also be available. Because the CR concept focuses on cost savings when it comes to checking invoices, unified billing is also a requirement of CR. Instead of checking separate bills for services from different providers, CR only has one monthly bill. Provisioning and operation of all services from only one service provider is not only a simple wish of the CR concept, rather it is more or less a consequence of the requirements listed earlier. To make unified service levels, unified support, unified billing, and so forth possible, single sourcing for all important infrastructure parts is the main precondition. Remaining in contracts with other suppliers compromises the CR concept.

If one now uses CR as a sourcing concept (and finds a provider who delivers and bills services that way) an interesting scenario surfaces: The bill for a single-source CR service is—at least at first sight—higher than the sum of the bills paid up to now to external providers for providing connections and services (see Figure 6.1).

But things look different when it comes to a TCO analysis (see Figure 6.2). Not only theoretical deliberations, but also experience with existing CR projects, show savings of about 20% to 30% compared with the total cost of what was previously necessary (costs for external providers as well as costs for assets, employees, and hidden costs).

With the characteristics described, CR differentiates itself from typical outsourcing and ASP offerings in these ways (see Figure 6.3):

- CR is a user-centered benefit-driven approach that puts the needs and wishes of companies buying services first.

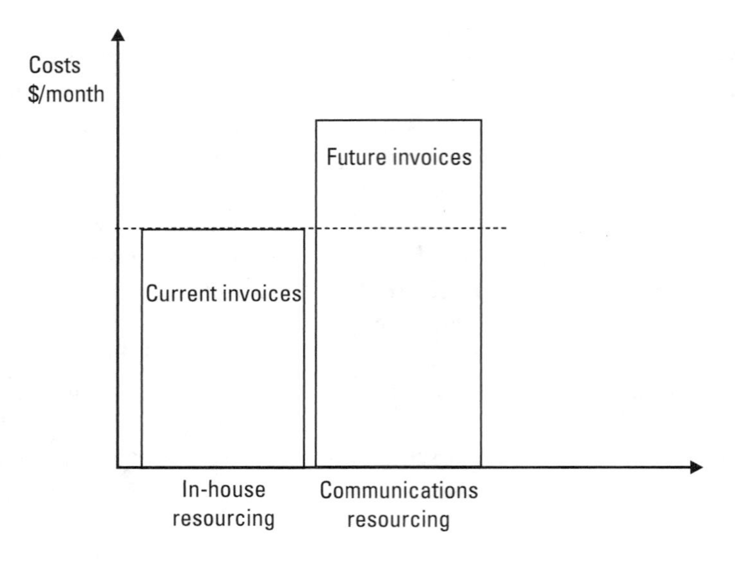

Figure 6.1 Development of the total costs for external services at the transition to communications resourcing.

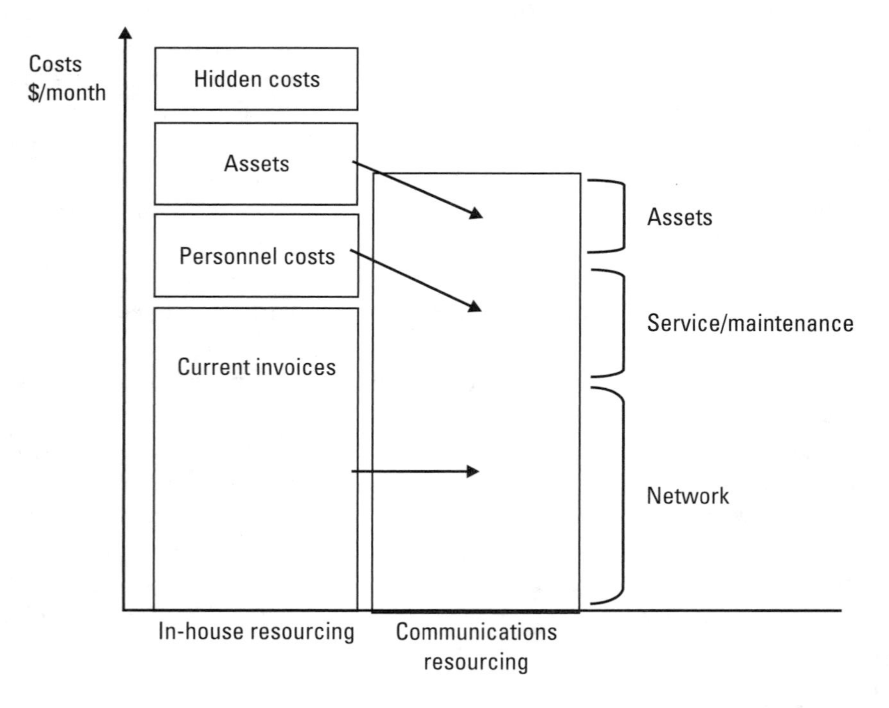

Figure 6.2 Development of the total costs' ownership at the transition to communications resourcing.

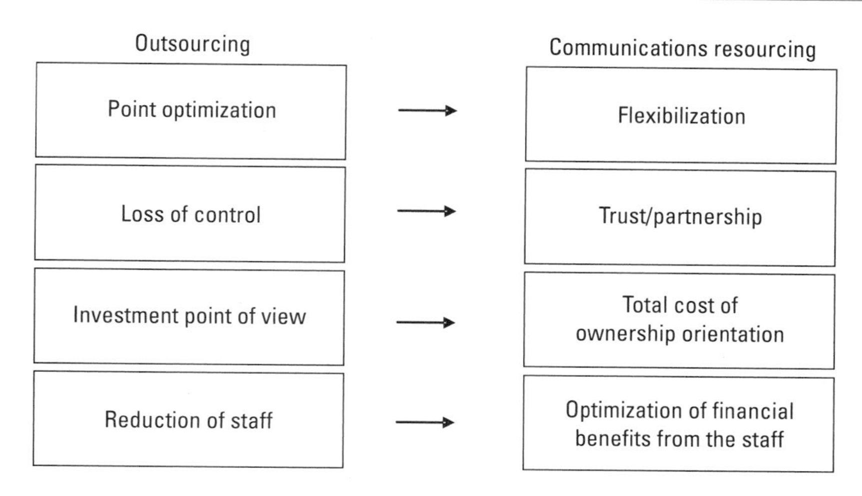

Figure 6.3 From outsourcing to resourcing.

- With its proven cost savings, CR is also relevant for industries where costs are a major issue, such as the retail industry.

6.2 Components

To fulfill the needs of enterprises, CR includes several service components. In the previous chapter, the phrase *provisioning of infrastructure for voice and data* was used to describe this. Of course, this is not very precise and needs to be explained. As one would expect, managed network services are a part of the concept, but not the entire solution.

Managed network services include the following:

- Network configuration and provisioning;
- Network management and operation;
- Support and service;
- Performance reporting and performance optimization;
- Network integration.

For WANs, CR integrates line offerings as well as the offerings of entire WAN infrastructures, whether shared or used solely by the customer. Typical service offerings end here, or are focused too much on technical parameters and less on business aspects. To build a complete offering from a customer's point of view, there are other services that need to be provided in this context. This of course includes typical hosted service offerings. Components that are useful here include the following:

- *Web hosting.* Web hosting services are usually purchased from an external provider so there is not much difference here.
- *Content management and e-commerce services* These are also examples of network-related services that are often externally provided.
- *Managed firewall/intrusion detection.* With growing Internet-related risks and dangers, and shortening time frames between security incidents and abuse by hackers, managed security services that can quickly react to changing environmental conditions are a useful addition.
- *Virus protection.* See preceding entry.
- *Security/authentication management.* This also plays a major part in a holistic view of the network.
- *E-mail/messaging.* This is the most important network-related service besides security.
- *Managed IP telephony.* This is an integral part of a converged infrastructure.
- *Storage and archiving.*
- *Conferencing and collaboration.*
- *Hosting of ERP software* (as far as it can be standardized).

With all of the services mentioned, significant cost savings can be achieved through economies of scale if the provider integrates several customers of the same service on its own infrastructure and service platforms. Self-provisioning would lead to significant higher costs here. Several of the services just listed are often already being purchased as external services instead of self-provisioned. CR now integrates and standardizes these services and integrates them with the basic networking services (WAN and LAN), and thus provides the basis for simple management and cost-saving provisioning and operation. This lays the foundation for a competitive CR model.

Analysts at the Yankee Group expressed a similar opinion in 2004 [1] (see Figure 6.4). They defined the term *productivity bundle* and differentiated three core components: (1) voice transmission, (2) broadband data transmission, and (3) network-related applications, which address security and productivity. The Yankee Group concluded that a combined sourcing of these three components provides productivity gains and potential cost savings. Yankee Group also mentioned that, in about half of all broadband contracts, management and security services are already part of the deal.

End user	Voice	Broadband data	Network-based applications	
			Security	Productivity
SME Medium-sized companies Large companies	Switched voice Mobile voice Intelligent numbers Web-based conferencing Hosted IP telephony	Internet access Managed broadband 3G/UMTS WLAN roaming IP VPN	Managed firewall Intrustion detection Antispam/antivirus Remote backup Secure messaging	Business applications Portals Hosted services PC support messaging
	1	2	3	

Figure 6.4 Productivity bundle according to the Yankee Group (2004).

6.3 Service Components

The success of any type of external service provision and outsourcing depends largely on the underlying service concept. A CR provider model must fulfill two basic requirements:

- Fulfillment of all corporate requirements to at least the same level of quality as that previously provided in a self-operated environment or from various providers;
- Transparent planning of service and support with the possibility of buying additional services and service extensions on demand.

6.3.1 Service Delivery

Assuming control of the corporate infrastructure also means assuming responsibility. To live up to the trust the company has in the provider, the provider must take further measures—apart from providing the core service—to build and support a trusting relationship. These measures include providing a dedicated "service manager" as a single point of contact to the service provider. This service manager is the person to speak to when questions arise regarding the network and all related applications. As a customer representative, he takes care of the needs and wishes of the customers.

Other measures include measurements that provide visibility of the provided services. The term *network visibility* is sometimes used for this. The provider proactively monitors the customer's network and provides her with a system for permanent surveillance of the system status. When network management is done proactively, all relevant parameters are permanently monitored by the provider and in the case of an incident the provider not only takes care of solving the problem, but also informs the customer on a regular basis (typically based on fixed time intervals) about the progress he is making in fixing the problem.

The additionally required customer service portal is no antithesis. Service not only lives from faith in proactive monitoring, but also needs to be proven every day by the provider. A main requirement of the CR concept is therefore a Web-based console in which all system management processes and trouble tickets are consolidated and displayed in a customer-oriented way. This console or Web application, often named a *customer service portal,* gives the necessary overview and also—if necessary—a detailed view of any part of the network. In an ideal world and in an ideal relationship between provider and company, the usage of this system declines over the contract period. After heavy use in the early stages (migration and thereafter), the system should be used less and less until nobody bothers anymore (that is, when everything always runs as it should).

6.3.2 Service-Level Agreements

Besides organizational and technical measurements for assurance and monitoring of operations, precisely defined SLAs are a necessary and important part of the CR concept. A single general SLA that includes all locations, components, network connections, and services lays the foundation for cooperation. Besides this general agreement, more detailed rules focus on single components of the infrastructure, depending on their importance.

SLAs should relate to business requirements, the perception of quality through the end user, and all technical requirements. In addition to this, the reporting system should focus on the needs and information wishes of the end user. The parameters defining the performance of the network—availability, latency, error rate, and error rate for connection initiation (dial-up)—must be matched to the parameters that are important from a user's point of view: availability and response times.

Availability is indeed a major factor. The current discussion relates mostly to the *x* in 99.*x* %. One hundred percent availability should be the universal goal when it comes to availability measurement. That this goal can only be approximately achieved by approximation is clear, but "100%" is more than the marketing hype of a network service provider (see Table 6.1). It is the final goal that should never be forgotten. Whether and how this percentage-oriented agreement is included in the master agreement is a question of business mathematics and differs from enterprise to enterprise.

Table 6.1
Availability Levels and Maximum Downtime per Year

Availability Percentage	Maximum Downtime Per Year (hours:minutes)
99.999%	0:05
99.99%	0:53
99.95%	4:23
99.9%	8:46
99.5%	43:48
99.0%	87:36 (more than 3.5 days!)

6.3.3 Integration of Other Suppliers

CR does not demand that all infrastructure-related services be done in every case by a single supplier. Due to industry-specific needs and legacy systems still in use everywhere, such a strict concept would not work. Industry-specific services, such as cash registers within retail companies or special software for computer-aided design (CAD) from industry-specific suppliers, should be integrated. The same is true of temporary services, for example, the continuing operation of an existing PABX/PBX system for a limited time during system migration. This places high requirements on the service provider.

6.3.4 Pricing Components

A major factor for CR pricing is a fair, transparent pricing model where every single component has its own price that is clearly understandable and reasonable. The main pricing model is the so-called "port" model. Besides this there is also usage-based pricing. Both of these types of pricing are discussed in the following sections.

Port Pricing

A *port* is defined as a technical functionality to support a business requirement, as well as a billing unit of a service provided within the CR concept. Here are some examples of port classifications:

- Data port;
- WLAN port;
- Telephony port a/b;
- Telephony port IP;
- WAN port.

Every port can be provided with a different set of features. The definition of the exact set of features for every port type is outlined in the individual agreement. An IP telephony port, for example, can sometimes include a special type of telephony device.

Depending on the individual needs of the project, it is possible to define ports for special requirements that are not included in the preceding listing. These can be ports for the connection of printing devices, barcode scanners, network cameras, and cash register systems or any other industry- or company-specific defined device that can be connected to the network.

Special comfort features can also be additionally supplied. Particularly when it comes to telephony ports, the features can be differentiated according to local requirements to match the expectations of the users, who are used to the traditional corporate telephony services.

The core of every service definition, therefore, should be a precise description of the different ports to be provided, and their features and requirements. The billing unit of the port model is a fixed price for every port and time unit (usually monthly). This billing unit includes all costs for provisioning and operation of the necessary infrastructure, as well as all costs for the necessary partial use of other network components, services, and devices.

Advantages of Port Pricing

The port-based pricing model has major advantages when it comes to infrastructure pricing:

- The costs per seat are clearly visible to the customer and remain stable for the entire contract duration.
- While reducing employee costs, no capital costs are necessary (variabilization of costs).
- Extensions, and also reductions, of the original number of seats can be provided almost immediately and lead directly to lower costs.
- Fixed costs, which are typical for self-deployed network infrastructures, do not show up for the end user. The provider takes this risk within its total service. Of course, the provider is interested in limiting this risk. Therefore, port models are usually limited to a specified fluctuation margin (for example, +10% to –10%). When the deviation is larger, the port prices are usually recalculated.

The port model provides flexibility and fulfills a basic need of IT/TC: variabilization and accountability. The advantages can be seen in, for example, VoIP installations: The variability of the number of ports is supported by another variability—the variability of locations—since VoIP allows easy migration of employees and their devices within the entire infrastructure by simply replugging the device into the new location.

If the port model is used throughout the entire infrastructure by also providing WAN ports, a precise estimation of the total cost of all intraorganizational networking can be made.

Example of a Port Definition

A general standardization of ports is not in sight. The service provider typically provides port definitions, unless the outsourcing company enforces its own specification. The following sample definition is useful in gaining a first impression of the necessary depth required to define a port description:

- *Connection bandwidth.* This is typically defined as 10/100/1000 Mbps. When it comes to data ports that are used for connecting printing devices, sometimes only 10 Mbps is fixed.
- *Availability.* Availability is described as a percentage. Apart from the numbers defined here, the location where availability is measured is important. When availability is only guaranteed at the port located within a network switch device, but not for the employee's desk, the entire cabling (the so-called "passive infrastructure") is not covered within the SLA and may also need to be serviced by a third-party provider if a failure occurs.
- *Redundancy.* Continuing operation of important parts of the infrastructure can be ensured by providing redundant connections. Normal data ports within the office would normally not be implemented and serviced in a redundant manner. Redundancy is usually defined as "no" when describing a typical port.
- *Management (proactive/reactive).* For a normal data port, management is normally done reactively. This means that, if an incident occurs, fault clearance only begins once the problem has been reported.
- *Parallel usage of voice and data (yes/no).* This depends on the underlying infrastructure. In "greenfield" installations, voice and data are usually supported in parallel.
- *Service level.* This includes the service parameters (operation time, service hours, response time, and fault clearing time) required to provide a concrete specification of a special port. For example:
  - *Operation time:* 24/7, with the exception of scheduled maintenance and repairs.
  - *Service hours:* Monday through Friday, 0800–2000, with the exception of country-wide public holidays.

- *Response time:* 3 hours within operation time.
- *Fault clearing time:* 8 hours within operation time.

The service level can, of course, be defined not only on a port basis, but also for all systems within a corporation or branch office.

It is absolutely necessary to give a clear definition of the port availability in a service specification. This is usually based on answer times that measure a signal between two devices on the network. This is very often done by simply sending data packets from the device to a default gateway and back. A sample definition of what availability of a port means is "A port is available when the answer time of 90% of all Internet Control Message Protocol (ICMP) echo packets sent is below 60 ms." It can also be useful to include measurement conditions and other limiting factors in the specification, for example, number of packets tested, packet size, system load, and so forth.

Billing by Usage

Some services cannot be billed by a port model alone. For example, CR cannot predict the quantity and duration of outgoing calls (to the public phone network), since one knows that the usage of services can depend on the pricing model used (e.g., flat-rate pricing leads to increased use and overuse, sometimes even abuse from a provider's point of view, in private surroundings as well as in business environments).

A flat-rate or port-based pricing system will be the exception when it comes to outside connections. When comparing network services with water and electrical power supplies, one can see that a flat rate would be absurd. If we had a flat rate for water supply with a designated bandwidth, water would pour out of the wall every time it was needed, too much for making a cup of tea but too little for filling the swimming pool within a reasonable time frame. A usage-based water supply (and billing) can lead to higher peaks than required by a continual stream and will be the better solution for most water customers. The availability of higher bandwidth when required will probably be worth more than a fixed price for an average bandwidth.

6.4 Benefits

Every development stage of CR provides clear benefits to the user. The core benefits include these:

- No capital lockup for network equipment;
- Reduction of operating costs;

- Simplification of the infrastructure;
- More flexibility in adoption to changes in business;
- Reduction of the TCO for the entire infrastructure (the total amount of the bill initially rises compared to what was previously spent on provider costs but then declines in comparison to the previous total costs of the infrastructure: bills for external services + employee costs + costs for assets + hidden costs);
- Increased security;
- Permanent update of the infrastructure;
- High reliability of the entire infrastructure.

The individual benefits for the person responsible for the corporate network infrastructure are "fewer worries" and "better sleep."

6.5 Communications Resourcing as a Process

For the success of the communications resourcing concept, not all components described earlier must be included, at least initially. CR can be seen as a step-by-step process or model. Already consolidating external connections can produce large cost savings. Further savings can be achieved by the integration of additional CR components, the change from traditional PABX/PBX to VoIP, and the use of network-related applications provided by the service provider.

6.5.1 Developmental Stages of Communications Resourcing

A step-by-step implementation can proceed as listed here:

- Consolidation of all wide-area connections and Internet access points;
- Consolidation of intranets, Internet, and LANs;
- Integration of VoIP within networks and applications (based on the port model described earlier);
- Implementation of network-related services and applications provided by the infrastructure supplier.

All development stages require the service model described earlier.

But a different approach, tailored to the special needs of a special company, is also possible. The steps just listed merely reflect a reasonably typical CR scenario.

6.5.2 External Provisioning and Operation of Network-Related Services

While the aspects and benefits of an integrated network infrastructure were described in detail earlier in this chapter, outsourcing of network-related services and applications within the communications resourcing concept requires additional description.

Network-related services and applications must be seen as an addition to the basic network infrastructure. Included are hosting services for Web applications and messaging but also operation of other network-related applications such as archiving and backup. These services are usually supplied on a common platform for all provider customers (as seen in the ASP model), but can also be serviced individually on a dedicated infrastructure to fit the special needs of the customer.

CR integrates the main benefits of on-demand computing by servicing applications in the provider data center that are delivered over the network infrastructure.

Apart from the economies of scale cost savings that can be achieved on the provider side, savings also accrue in capital expenses for the customer. Usage-oriented and seat-based pricing helps to provide a cost overview and to allocated the costs to the place where they occur. Even small user bases can be served with reasonable costs and a defined service level. For a higher number of users, the services can also be individualized within the CRconcept. This can then also include the provision of high-level services such as ERP.

To sum this up, one can say that purchasing applications via the CR concept has the following benefits:

- Cost optimization due to application usage oriented to actual needs;
- Little or no capital expenditure;
- Cost control/high transparency and easy clearing;
- Fast and flexible adaptation to changing needs and environmental conditions;
- Simple but effective self-service;
- Lowering of TCO;
- Increased security due to redundant data storage within the data center and active responses to security threats.

CR includes components and ideas of what is elsewhere called *on-demand computing*, not as the core element but only as an additional element of the CR concept.

6.6 Case Studies: Communications Resourcing at Work

CRis no vague business concept but an already-proven approach to external purchasing of network infrastructure and related services. The following case studies of international companies show the way. The list is in alphabetical order and implies no ranking. All companies listed are involved in a CR process. The paths taken are not the same, but the direction is clear.

6.6.1 Amway

Founded in 1959, U.S. company Amway (Ada, Michigan) is one of the largest direct marketing companies worldwide (see Figure 6.5). About 6,000 employees, and more than 3 million resellers, work for Amway worldwide. They sell more than 450 different products in more than 50 countries. The products, solely sold direct to the customer, include nutrition, wellness, cosmetics, and household goods.

Status Quo and Needs

The strongly growing sales network and strongly increasing revenues of the European organization of Amway, which includes 23 countries, made consolidation of ordering processes necessary. Amway was an early adopter of online ordering, for example, with BTX (Bildschirmtext, a proprietary wide-area information system, similar to CompuServe or early AOL, started in the 1980s in Germany and in a similar form throughout Europe), and also with one of the first e-commerce ordering platforms in Great Britain.

The main goal was improved sales processes with the best Web infrastructure in its class for around-the-clock order entry and forwarding.

Requirements

The following requirements were initially defined:

- Unified Web portal with online ordering for registered resellers;
- Consolidation of the infrastructure of 23 country branches;
- Integrated infrastructure for Web hosting and network services, concentrated at one partner (since experiences from the past made interface problems obvious);
- High availability (99.5% or better), since all solutions used previously had trouble achieving a sufficient level of quality;
- A maximum response time of 1 hour;

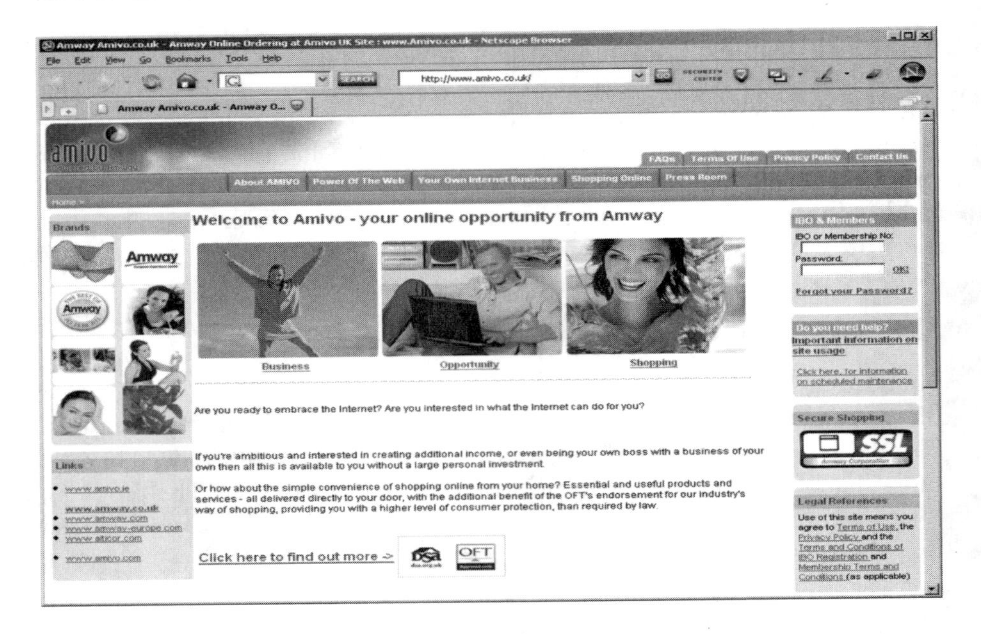

Figure 6.5 Amivo online order system.

- Simplified hosting and operation of the entire system;
- Cost savings.

Solution

What was special about this solution is that the main challenge was to unify everything into a singe Web portal and online order system as the crucial part of the sales chain.

Amway decided to buy its service as a communications resourcing solution from T-Systems international. From a technical point of view, the solution consists of 22 server systems, which are provided in the T-Systems hosting center in Magdeburg, Germany. This server infrastructure is connected via an MPLS network with the European headquarters in Puchheim (near Munich), Germany.

Sixteen servers are used within the data center for the Amway portal system (see Figure 6.6). Four IBM B50 AIX systems are used for load balancing as an EDGE cluster. Four other IBM B50s are used as Web servers, while four IBM pSeries 640 systems are used as application servers (within a server cluster). The software solution is based on IBM's Websphere Application server. There also is a DB2 Server cluster, which uses two series 640 IBM machines with FibreChannel. The staging area includes one DB2 Server cluster, two application servers, one Web server, one MQ server, and one EDGE server. The system

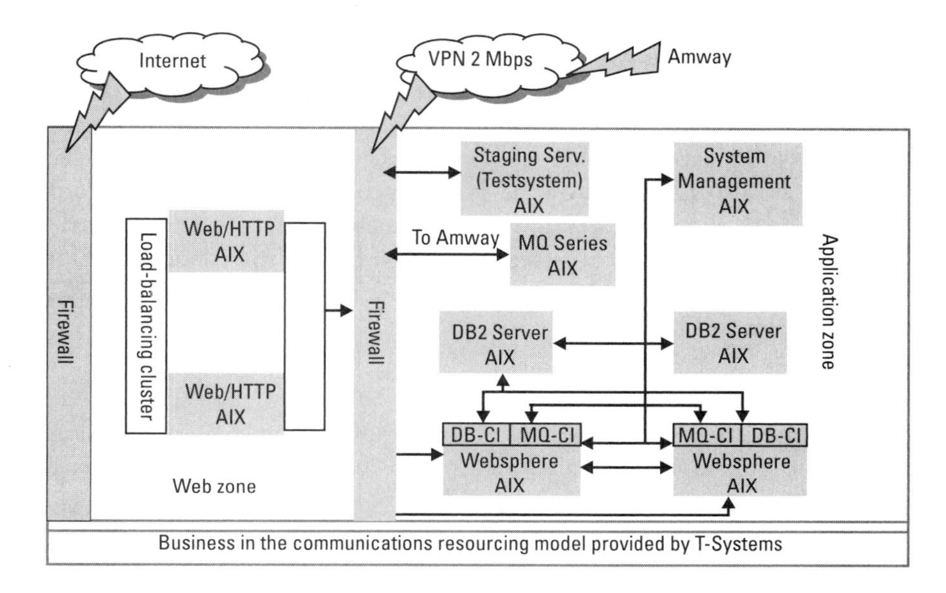

Figure 6.6 The Amivo architecture.

works with incoming data and transfers it in online mode per MQ server to the European data center in Puchheim. If there is a connection loss, the front end can continue working without losing orders. As soon as the connection is working again, the system automatically transfers the data to the back end.

The new solution offers several advantages over the historic infrastructure:

- A unified, centralized network platform for order entry and order acceptance of all orders generated by the resellers;
- Increased quality, scalability, and security;

Guaranteed availability with clearly defined service levels (99.5 % availability, 1 hour response time), scalability, and high security;

Extension of the Web-based business; besides order entry, the system also integrates availability checks, downloads product information, and searches for regional partners;

- Massive cost savings through centralization and consolidation of systems that were previously country specific.

6.6.2 Fossil

Fossil Inc. is a worldwide company, based in Richardson, Texas, that offers watches, jewelry, and leather goods for sale (see Figure 6.7). Fossil has more than

30 sales branch offices, logistic centers, and design facilities worldwide, including Canada, Australia, Japan, New Zealand, China, Belgium, Spain, France, United Kingdom, Italy, the Netherlands, and Germany.

Fossils product offerings are sold in more than 90 countries via retail channels. There are also 120 Fossil-owned stores, Web sites, and e-commerce order systems for resellers and end customers. The main data centers are located in Richardson, Texas; Erlstadt/Eggstaett, Germany; Basel, Switzerland; and Hong Kong. More information on Fossil can be found in their global corporate Web site (http://www.fossil.com).

Status Quo and Needs

Before Fossil started realignment of their worldwide network infrastructure, there was no unified network platform and no "one-stop" service provider for the operational requirements in the United States and internationally.

Challenges

Fossil was in search of a partner that could offer an international MPLS-based platform and provide the entire WAN. The partner also had to bring knowledge in ERP system (SAP) rollout.

Figure 6.7 International Fossil Inc. Web site.

Solution

Fossil migrated its WAN to a multinational network provided by T-Systems International. The solution provides a cost-effective and flexible business platform and provides various advantages and improvements including network operation, business operation, and customer relations.

When it comes to network operation, the MPLS platform allows different service classes (at Fossil the following are defined: voice, business, and economy), flexible service features at every location, the ability to track trouble tickets, project implementation and rollout, and network performance via a Web portal. Also, when it comes to business applications, Fossil focuses nationally as well as internationally on one single strong partner. For the end users this translates to better response times as well as improved service quality.

6.6.3 OMV

With more than 6,000 employees and around 7.6 billion Euro revenue in 2003, OMV is one of the leading oil and gas companies in Central and Eastern Europe (see Figure 6.8). Business activities include worldwide search, exploration, and production of oil and gas; oil and gas supply; and a chain of gas stations throughout Central and Eastern Europe. With its expansion-oriented strategy,

Figure 6.8 OMV company Web site.

OMV buys or holds shares in companies such as MOL (Hungary) and the Rompetrol Group (Romania).

OMV's strategic goal is to double revenue by 2008 with an increase in capital profitability to 13%. With the corporate headquarters located in Vienna, Austria, and a strong understanding of Central and Eastern European markets, as well as a highly motivated staff, the chances of achieving this are high.

Status Quo and Needs

To reach these lofty goals, the telecommunications infrastructure must grow with the increasing requirements, or even pave the way for expansion.

Back in the 1990s, the OMV company network was limited to their home country of Austria. At this time, the connection to local production facilities was the top priority.

Since 2000, an international frame relay network has been constructed step by step, largely as a result of the new requirements that came with implementing a SAP solution. This also led to the implementation of other network-related applications (such as a unified messaging system based on Microsoft Exchange and the OMV intranet portal), and the connection of even more locations.

For the locations in Austria, Germany, Hungary, Czechoslovakia, and Slovenia, all applications were centralized and supported by a unified management structure. With the acquisition of several hundred gas stations in Germany in 2003, the company-owned sales network grew strongly. OMV currently has more than twice the number of outlets outside its home country (13 European countries) as in its home country. This must all be supported and connected.

Requirements

The main requirement was building a unified, strong network infrastructure for an international, strongly growing company with branch offices and production facilities in locations one could easily describe as exotic. This is not an easy challenge, especially when it came to finding the right partner and contracts, since there were different views on the topic throughout the company.

With the new network, the frame relay system was to be discarded, since OMV viewed the technology as a dead-end road. Of course, cost savings were also a major issue—as everywhere. The goal here was to limit the budget to previous expenditures while also providing more and more bandwidth at the same cost. Of course, OMV knew that additional cost would be necessary for the connection of additional locations, but for the core network the following idea was laid down: Obtaining more every year without an increasing price tag.

Besides pricing discussion, further expansion needed to be integrated into the contract. So a flexible contract supporting the expected change requests was a major factor. The following criteria were defined for the selection of the service provider:

- *Proven technological competency.* Proven competency in the technology was required for complex infrastructures.
- *Internationality.* The provider should be able to provide its own resources within the most important countries, and also offer local competency and local language support for the remaining countries.
- *Industry competence.* The partner should have experience with and an understanding of the oil and gas business.
- *Trust, efficiency, and flexibility.* Strong customer orientation, continuous service in sales as well service and support ("one face to the customer"), and a transparent organization on the supplier's side were required.
- *Cost/offering.* The provider's costing needed to be transparent and it needed to be able to supply the required level of quality.

Challenges

The most important challenge was the strong internationalization, especially when it came to exploration and production. Exploration not only occurs in Europe, Africa and the near and middle east but also in South America and Australia / New Zealand, while marketing and sales focuses on Europe and Eastern Europe.

To avoid interface problems, the solution was to be provided by one service partner only.

Solution

OMV contracted with T-Systems International for its WAN. The existing infrastructure was migrated to MPLS technology. The sales manager of T-Systems is located in Vienna, as are the OMV headquarters. The network grows and additional locations are connected as needed. Development is controlled by the OMV daughter company, OMV Solutions GesmbH, which delivers services to OMV that can be centralized, such as human resources, medical services, and facility management. OMV solutions rely largely on partners, not only when it comes to networking, but also in other service areas. About 75% of all services are externally purchased.

The provider does network management proactively. OMV has its own Web-based console and has a detailed overview over the entire network available at any time.

The new provider-based OMV WAN now also forms the foundation of the centralization of other services such as messaging, directory services, storage, and software distribution, updating, and patch management. The standardization that can be achieved with this has led to additional savings for OMV.

The agreement between OMV and the service provider includes a best-price condition. The goal described earlier supports obtaining more bandwidth with the same budget. By changing the network to the new service provider, costs are down about 10%, while bandwidth has increased by a factor of 3.

Future

MPLS technology used in the OMV WAN delivers the foundation for the integration of voice services, which could be a future step in improving the network. Because there are existing contracts with PABX/PBX suppliers, this is still in the planning stage, but extensive testing and the use of VoIP in certain areas (such as call centers) broadens the way to a future voice implementation. There are also other locations that need to be connected in the future.

Cost effectiveness and high flexibility fit together well at OMV. CR helps OMV to support future growth with a growing, adaptive infrastructure.

6.6.4 Smart

Smart GmbH, located in Boeblingen (near Stuttgart), Germany, is a 100% daughter company of DaimlerChrysler and a pioneer in new vehicle concepts (see Figure 6.9). The second generation of what is known as "Smart ForTwo," a small (only 2.60-m-long) two-seater compact car, will hit the U.S. market in 2006 or 2007. The first generation was introduced in Europe and has been sold in several export markets since 1997. While there are now other models—a two-seater compact roadster as well as a more standard four-door compact car built together in a joint venture with Mitsubishi in the Netherlands—the original Smart concept was, and still is, the "ForTwo." It is built in an innovative factory in Hambach, France, with engines built in Koelleda, Germany. Because suppliers do most of the work—even within the plant—there is only an 18% share of work that is directly done by Smart workers—following the lean factory concept.

Costs were not the only relevant issue in choosing a new partner for the corporate LAN in 1999, but also a shortage of personnel and the impending Y2K (year 2000) problem. The idea was to obtain a more flexible solution that supported the expected growth while also supporting DaimlerChrysler standards and cost targets.

The solution found was simple. Smart buys its LAN services based on a port model and other services and system integration from T-Systems International. About 3,000 LAN ports are currently managed within this structure. For

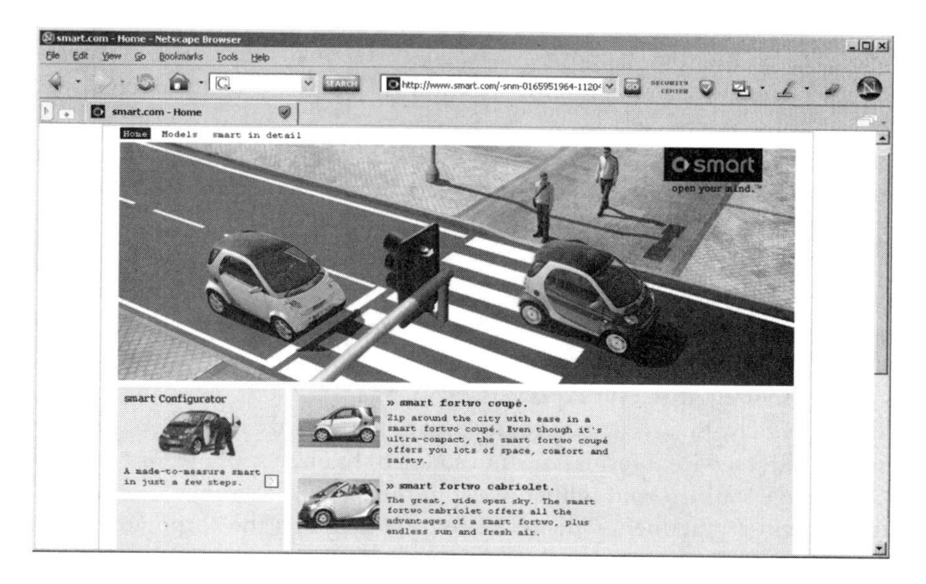

Figure 6.9 Smart company Web site.

Smart this means significant cost savings, increased flexibility, and simple scalability.

Services also include WAN connection to about 300 sales outlets via xDSL (supported by ISDN backup). The services are centralized and can now be used via Citrix. Also, e-mail service and Internet access are provided to the entire dealer network from a single point. In addition, the infrastructure is provisioned, operated, and serviced by T-Systems International. Reduced complexity within the infrastructure and better application availability help Smart to support the end user's wishes and needs. Now more than ever car, sales need to be flexible to support customer wishes such as precise information on delivery times.

6.6.5 UC4

UC4 Software in Vienna, Austria, is one of the world's top four players in data center automation software. Their main software product is a solution for complex tasks in job scheduling and event automation. UC4 has nine major locations worldwide. Besides Austria, there are locations in the United States (Denver, New York), United Kingdom (London), Germany (Munich, Frankfurt, Braunschweig), Switzerland (Zurich), and Australia (Sydney) where about 130 employees work for more than 600 current customers. Most customers are large multinational enterprises, so UC4 already has more than 200,000 installed licenses. User support is done worldwide in a "follow-the-sun" concept, via a single toll-free number supported in English, German, French, and Spanish. Customers include such prominent worldwide players as DaimlerChrysler, Siemens, Bosch, Henkel, Cadbury Schweppes, and Austrian Airlines. The unique selling proposition of UC4 Software is this: It supports almost every system in use anywhere in the world, particularly legacy systems.

Status Quo and Needs

As an innovative software maker, UC4 sees its IT/TC strategy as a major part of its corporate strategy. To connect the worldwide branch offices with each other and with the headquarters, UC4 made early use of frame relay technology. All data is centralized at the headquarters in Vienna. In 2003, the main corporate Web site, which also acts as a customer service portal and includes not only content (hot fixes, service information, and so forth) but also support functionalities such as usage statistics and billing, was outsourced to T-Systems International. But the main requirement when it came to redesigning the corporate network was to also provide 24-hour worldwide customer support.

Requirements

Problems and inadequacies with the frame relay infrastructure led to the search for a new solution for UC4's WAN infrastructure. The frame relay infrastructure supported no QoS classes, and was constructed in a star formation and not as a mesh as was needed. Due to cost reasons, the U.S., U.K., and Australian branch offices were not connected to UC4's frame relay infrastructure. Also, service-level guarantees were only available for the frame relay network itself and not for the local loop. Change costs were high. To sum this up, one can say that the situation did meet the expectations of UC4.

The consequence was simple and clear. With the original frame relay contract running out in April 2004, it was necessary to provide a new solution without compromising cost and flexibility.

When selecting the vendor, the following requirements were most important:

- 24/7 operation;
- Very high availability;
- High flexibility for making changes (new branches, migration of branches, changes in bandwidth used).

Solution

UC4 now uses an MPLS data network provided by T-Systems. The existing infrastructure was migrated to MPLS when changing providers. All connection bandwidths were increased. Up to four times more speed is available now within the UC4 WAN.

The change to MPLS made differentiation/prioritization of service classes available as well as any-to-any connections. All branch offices, except the one in Australia, are now integrated into the new WAN infrastructure. The SLAs are now available for the entire network and are no longer limited to the network backbone. The infrastructure is now supported by ISDN backup, and the headquarters now has redundant connections to the outside world. Also, Internet access was concentrated. All European Internet traffic is routed via Vienna, while all U.S. traffic is routed via the firewall in Denver, Colorado.

The total cost savings achieved were about 15%. Specific savings reached up to 30% compared to the old solution when examining single connections or services. The most important factor speaking for the new solution is increased flexibility. This helps when changes in the infrastructure become necessary, for example, due to new locations, location closures, and bandwidth changes.

Future

UC4 has laid the foundation for further strong growth with the new definition of its corporate WAN. While VoIP plays no major role in its planning (there is little voice traffic between the locations), UC4 currently runs tests at its Frankfurt location. Flexibility will be the major issue when it comes to future growth.

6.6.6 Umdasch Doka

Umdasch Corporation is a global player with branch offices and daughter companies in 35 countries (see Figure 6.10). With its two segments (formwork and shop fitting), the Umdasch group of companies reached a turnover of $616 million in 2004. With more than 4,800 employees, the company is a leading international player.

The Umdasch headquarters are located in Amstetten, Austria. Umdasch shop fitting is one of the best-known shop fitters in the food and nonfood sectors throughout Europe. Doka is an international brand in formwork and supports construction companies worldwide with innovative products and services, for building better, faster, and cheaper.

Status Quo and Needs

More than 2,200 devices and 141 server systems were located at 99 locations in 35 countries. This very decentralized structure suffered severe problems. A multitude of interfaces were normally connected with only asynchronous data communication. Support became increasingly difficult, since there was no local IT competency, and no IT employees were available within the branch offices. In the headquarters, no three-shift service available, due to economic reasons. Such

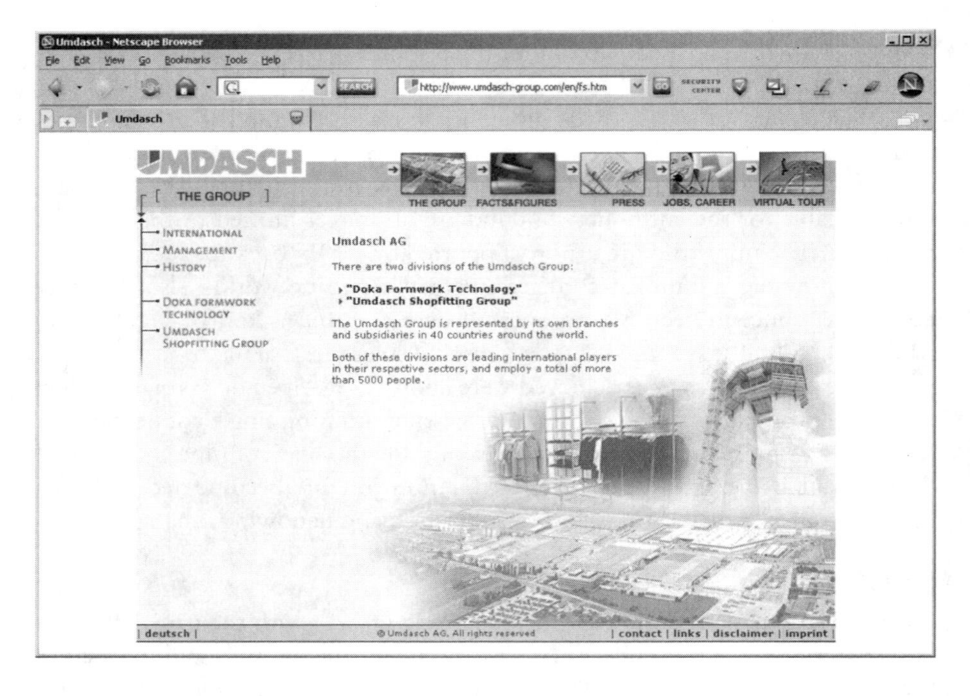

Figure 6.10 Umdasch group Web site.

service was, however, necessary in order to support construction work in other time zones.

Besides the Austrian headquarters, only one branch had, and still has, a strong standing: Maisach in Germany, connected via frame relay. Most other locations were loosely connected via ISDN to one of the two major locations. The consequences for Umdasch Doka were two networks and difficult peer-to-peer access that was only achievable with several workarounds. Because both basic structures were entirely different (for historical reasons) a simple consolidation approach would not have worked.

The main disadvantages of the old infrastructure were as follows:

- ISDN connections were cost intensive and made it difficult to calculate the cost for different locations.
- The infrastructure was not sufficient for the general plan of centralizing services.
- The VPN system used was no longer supported by the manufacturer and needed replacing.

Requirements

The main goal was the standardization of the entire infrastructure through the elimination of interfaces. Besides this, the problems relating to multiple providers needed to be reduced. The best way to reach these goals—from Umdasch's point of view—was to outsource the network operation to one reliable partner. Even in the early planning phases, SLAs were a major factor.

The following goals were defined within the planning process:

- Better service quality;
- Transparency of costs and relations to users;
- Retaining of strategic competency at Umdasch;
- Centralized contract management at the headquarters;
- 24/7 service (necessary for a global player; Umdasch was not able to provide this service on its own).

To make the supplier selection process controllable, Umdasch decided on an internal analysis of the status quo (analyses of its own network and of technologies available on the market) and development of a system specification with the help of an external consultant. The conclusion was that leased lines for all locations would have been possible but not cost effective. Apart from what was seen as the best solution (MPLS), basic IP VPN was also considered.

Seven providers were evaluated in the bidding process. Four offers were received but not all of them were as expected and did not comply with the listed specifications. After this initial run, there were discussions with two suppliers. An agreement was then made with the supplier defining an SLA, a framework, and final pricing.

Solution

Note that when selecting a new provider, the one initially selected went bankrupt during agreement finalization so the bidding process had to be repeated.

The new Umdasch group WAN is delivered by T-Systems International as a CR project with the following conditions:

- MPLS VPN service for the major branch offices (bandwidth between 128 KB and 4 MB);
- Multiple connections for the major locations of Amstetten and Maisach;
- ISDN backup for the most important locations in Germany, Switzerland, and Slovakia;

- "Exotic" locations connected via the Internet gateway service offered by the service provider;
- For security reasons, connection to foreign networks and mobile users via their own VPN gateway;
- Concentration of Internet access at the two major locations with their own 4-Mbps connection.

The solution is provided on a 24/7 basis, including monitoring and help desk services, under the SLAs. Umdasch has an overview of the network activity available at any time since they have access to an individualized Web-based customer service portal. All data traffic is logged in the headquarters to complete the overview.

The solution has established itself as a stable platform for IT consolidation. Higher system stability and higher speeds compared with the previous situation have led to remarkable gains in productivity.

Applications such as SAP, document management, and directory services are now provided centrally. With the new solution, an employee can easily temporarily work somewhere else within the WAN using the same environment she is used to. This was close to impossible in the old infrastructure. Cost savings can also be achieved by centralizing license keys, so that concurrent CAD users can be limited.

The benefits of MPLS worked out well. During the bidding process, everyone thought that MPLS would be merely "nice to have," but the benefits can now be reaped, especially when it comes to functionality that uses the mesh structure the system now offers.

Future

Although there is skepticism about the financial benefits of VoIP, an internal business case is currently in the works. However, the first priority is, and will be, the connection to other branches, such as the Near and Middle East, the Baltic, South Africa, and Bulgaria. The flexibility of CR pays off.

6.7 The Corporate Network of the Future

Communications resourcing has proven its strength. The examples given in the preceding section are just momentary snapshots of individual processes, but they show the potential of the CR concept for very different companies and business models, ranging from oil production to software design.

The dream of turning voice and data infrastructures into a commodity, in the same way as for water and electrical power supplies, has not yet come true.

The differences in company requirements for service and availability are too disparate to make this happen any time in the near future.

The most important step toward drastic simplification, higher flexibility, better quality, and TCO savings can already be delivered using the CR concept. This includes, but is not limited to, the port model as a major anchor for a new orientation in buying voice and data services for corporate use.

What will the future of communications resourcing bring? Several questions are relevant for the future development of CR that must remain open here:

What will happen to fixed/mobile integration in the context of CR?

Is there a way of integrating highly individualized application delivery?

What service providers offer CR or similar concepts?

How can small companies profit from CR?

The scope of this book does not allow these issues to be addressed as yet. See http://www.communicationsresourcing.de for more information (in English and German).

Reference

[1] "Productivity Bundle," Yankee Group, 2004.

Glossary

2G network Mobile phone networks are usually split into different phases or generations. GSM networks are included in 2G networks (second generation networks). (*See also* GSM.)

3G network Networks of the third generation are usually UMTS networks (*See also* UMTS.)

Active-X Active-X describes a technology, developed by Microsoft, that software components in a heterogeneous environment can use to communicate with each other. Active-X is built on top of the Common Object Model (COM), which defines a standard for communicating between objects. Active-X is mostly used for creating interactive elements in Web applications.

Always-on This is a constant connection to a network or service, typically with the end device. An always-on connection keeps an established connection open, even when it is only infrequently used, and saves the tedious process of reconnection. Incoming messages are passed to the end device without delay instead of having to wait until the next time the user makes a connection. The increasing replacement of dial-up connections with permanent connections, especially DSL-based wideband connections, makes it much easier to use company applications anywhere. Always-on is also possible over mobile telecommunications using packet services such as GPRS, EDGE, and UMTS.

American Standard Code for Information Interchange (ASCII) This is an ISO standard code for representing the alphabet, including large and small letters, punctuation, and special characters.

Any-to-any This describes the ability of establishing a communications link between any two members of a network.

Asynchronous Transfer Mode (ATM) ATM is a base technology for WANs that allows the transfer of transmission-critical applications. The ATM technology, which lies in the two lowest layers of the OSI architecture, is based on the storage forwarded transfer of data in the form of asynchronous, addressed, fixed-length cells (cell switching). ATM systems are connection oriented, and the end devices connected to the network communicate over virtual end-to-end connections. The concepts of virtual channel (VC) and virtual path (VP) are used to describe the path of a virtual connection over an ATM network. All cells for a particular connection at a defined speed (bit rate) follow the same path through the network.

Authentication management Management of authentication.

Backbone A somewhat inconsistently used term for describing the "spine" of a network infrastructure (i.e., a high bandwidth connection between the various subnets in a large network).

Base connection *See* ISDN.

Benchmark Benchmarks are test processes that form a basis for comparing the performance of different technical devices or organizations.

Best-effort delivery Describes the transfer of data without assured quality of service.

Best-price clause A contractual agreement in outsourcing contracts that allows adjustment of the price if market prices fall. Defining the reference value is usually somewhat problematic.

Border Gateway Protocol (BGP) Used to define how routers propagate information about the availability of network paths to each other. The currently used version of the Border Gateway Protocol is described in RFC 1771 (http://www.faqs.org/rfcs/rfc1771.html).

Broadband; broadband connection; broadband access This term is not always used consistently. Typically, this describes end-user Internet access systems having a significantly higher bandwidth than a dial-up modem or ISDN, for example, based on DSL or a cable modem.

Broadcast The simultaneous transmission of information from a single point to all other participants. Typical broadcast applications are radio and television. A data network also offers the possibility of simultaneously communicating with a particular class of recipients, a particular subset of the network (multicast), or all connected systems (broadcast).

Buffer A place where data is temporarily stored if it cannot be immediately sent over the network. Buffering can be an important factor in the transmission of audio and video data over a network.

Business process outsourcing (BPO) The transfer of complete business processes to a third-party company.

Business transformation outsourcing (BTO) A process in which business-critical processes are taken over by a service provider, optimized in accordance with the business strategy, and, if desired, reintegrated into the company in an improved competitive state once the contract is finished. The main focus here lies in process improvement.

Business TV The process of video transmission for closed groups of users, usually within a company or organization. Business TV is mainly used for informing users, or for further education of the employees, and is increasingly available in intranet systems. Other definitions of business TV include the possibility of providing content for suppliers, which is not always without problems because of the legal questions that it raises (e.g., with regard to licensing).

Business-to-business (B2B) A business relationship between two businesses, in contrast to business-to-consumer (B2C).

Carrier Originally used in the telecommunications industry to describe a high-frequency signal that can be transmitted via an antenna or cable. In cable communication, a carrier is an AC signal in the cable. In optical communication, it is an optical signal, that is, light of a specified wavelength. In the context of this book, it is also a company offering real or virtual cable connections.

Category 3/Category 5 (Cat-3/Cat-5) Cabling standards for buildings. They are defined according to possible application and bandwidth (see Table G.1).

Table G.1
Comparison of Cabling Standards

Component Category	Network Application Class for Point-to-Point Connections	Defines Frequency (MHz)	Application
Category 3	Class C	16	Low-bit-rate speech and data transfer
Category 5	Class D	100	Speech and data transfer

CENTREX Acronym derived from Central Office Exchange Service; relates to the extended use of a public network node for the provision of a virtual telephone exchange.

Change request Describes a customer request for a change or improvement to a system that is not contained within the scope of the agreement or is differently defined. Payment for change requests is often the cause of conflict between customer and service provider.

Chinese walls These are information barriers, existing within a financial institution, used to avoid a conflict of interest by preventing individual departments from accessing information about the actual activities of other departments.

Citrix Citrix is a supplier of Citrix Metaframe, a system for application distribution and management. This functions on a service-centered basis that primarily transfers only screen content to the client and receives input from them.

Clarity A characteristic of quality in telephony systems.

Class of service (CoS) This is a classification characteristic for handling data packets according to particular priority rules. The traffic can be prioritized in the following order: speech, video, ERP applications, Internet use, and e-mail/downloads.

Client; client-server Corporate computer networks can usually be categorized by the type of end devices that they use: clients and servers. Clients are usually the computer at the desk of the employee. These then utilize services provided by a server. The tasks performed by clients and servers can vary, depending on the type of client-server model used, for example, the location where data is stored.

Compression/decompression (codec) An invented word created from *compression* and *decompression.*A codec is a data transformation unit that modifies audio/video signals in real time according to a fixed process. The data transformation unit can be implemented purely in software but can also be a dedicated hardware unit. The actual transformation processes used by different codecs are standardized by the ITU and described in various ITU recommendations. The various different compression algorithms used can cause very large differences in the sound and picture quality of the processed signal.

Commodity Describes standardized and, therefore, exchangeable products and services.

Communications resourcing A new concept for optimizing the structure of third-party provision of speech and data networks as well as network-related services and applications.

Company portal A central application in the company, usually implemented using the WWW standard. The main task of a portal is to offer the use a single point of access to various company information and services. A portal prepares this information for the user in a common format. Company portals usually also offer typical security mechanisms and other additional features to assist the user, such as centralized log-in, search systems, and directory functions.

Competitive intelligence An element of modern business leadership that has the aim of achieving competitive advantages through the acquisition, and systematic analysis, of specially targeted market and competition information.

Content management system (CMS) Software for managing content. It is usually synonymous with software used for managing Web content (WCMS = Web content management system), but is also often used to describe a type of editorial system. The actual meaning extends much further than Web applications and increasingly describes systems having document management functionality.

Convergence The growing together of two or more platforms or technologies.

Corporate espionage Process of obtaining unauthorized access to company-internal information.

Corporate network A business network conceived for special requirements. The network can cover a specific location, a region, a country, multiple countries, or possibly the whole world, depending on the company structure. A corporate network can include the entire range of telecommunications services, for example, the transmission of data and speech, but also added value and messaging services such as e-mail.

Corporate Web site The official Web site of a company.

CPU usage The level of usage of the main processing unit in a computer.

Customer premises equipment (CPE) Hardware required for implementing a network that is installed on the customer premises.

Customer satisfaction An index of how happy the customer is. Improves the SLA concept by continuous, or repeated, evaluation of customer satisfaction. The payment of premiums is often based on achieving specific levels on a customer satisfaction index.

Database management system (DBMS) An administration system for databases. Describes programs that allow permanent, application-independent storage of data in a database and administration of this database. Other tasks of a DBMS are: preparation of various views of the data, organized in different ways (also known as "views"); checking the consistency of the data (integrity checking); authorization checking; handling of simultaneous access by different users (synchronization); and preparation of data backup possibilities in case of a system failure. A DBMS is often confusingly referred to as simply a *database.*

Delay In communications technology, the period of time the signal takes to arrive at the receiver. Conversion and network-specific delays are unavoidable when transmitting speech over a packet-forwarding network. These occur in the audio codec, in the packet creation process at the sender, and during the buffering process in the network. Delay affects the perceived quality of a telephone connection but is not relevant in single direction connections (radio model)

Desktop management/desktop services The provision and support of workplace hardware and software. This term is normally used in the context of outsourcing.

Device connection *See* ISDN.

Digital Enhanced Cordless Telecommunications (**DECT**) An ETSI standard for cordless telephones and also for general wireless communication. DECT is primarily designed for so-called picocellular telephony within buildings. Ranges of up to 50m within a building and 300m outside a building are possible. The radius of operation can be extended through the use of repeaters.

Digital Subscriber Line (**DSL**) DSL can be summarized as a range of processes allowing wideband use of normal copper wire telephone cables between the subscriber and the telephone exchange. The most common variant is ADSL (Asymmetric Digital Subscriber Line). In this case, there is a difference between the upstream and downstream transfer rates that can be achieved. In SDSL, the bandwidth is identical in both directions. xDSL is often used as a generic term for these and other DSL variants. DSL/xDSL can be regarded as one of the more important end-user wideband access technologies and allows low-cost connection of satellite companies and branch offices to the company network.

Disaster recovery A process by which systems are placed back into productive operation after a serious breakdown in operations with data loss has occurred.

Disruptive technology The new technology that destroys, or significantly impairs, an existing business model or strategy. For example, consider the effect of the development and marketing breakthrough of digital cameras on traditional film manufacturers.

Downstream Describes an incoming flow of data. In IT, one can describe a flow of data as a sequence of data where the end of the sequence cannot be predicted in advance. (*See also* Upstream.) The upstream and downstream bandwidth of a connection do not need to be identical. Mobile data services, and also consumer-oriented Internet connections, often provide different bandwidths for upstream and downstream traffic.

Echo Echo is the reflection of the transferred signal back to the sending device. Echoes can have a negative effect on the speech quality.

Electronic business (e-business) Describes business processes that are performed electronically. The use of the term has been popularized by the service providers and is not very clearly defined.

Electronic commerce (e-commerce) Trading of products and services using electronic means for communication.

E-mail Electronic mail.

E-mail push A process for distributing electronic mail to mobile devices. In contrast to normal systems, which require a request from the client device, e-mails are automatically forwarded to the client device as soon as the e-mail server receives them.

Enhanced data rates for GSM evolution (EDGE) EDGE describes a technology for increasing the data rate in GSM mobile radio networks. Like GPRS, EDGE is an evolutionary development of the GSM technology that can be built into the mobile radio network with relatively little effort (mainly software updates in the sending device and exchange of a few components) and does not affect the functionality of existing mobile phones. The data rate is increased to approximately 48 Kbps per channel per user (up to 384 Kbps using eight channels). This is achieved by changing to a more efficient modulation process (8-PSK) This change occurs selectively only on the channels used by EDGE-capable devices. This allows simultaneous trouble-free use of GSM/GPRS and EDGE-capable devices on the same network. EDGE is regarded as an intermediate technology on the path toward UMTS. Up to now, only a few providers have adopted EDGE.

Enterprise resource planning (ERP) A company task in which the resources in a company (e.g., capital, operational resources, personnel) are planned in the most efficient manner for company operation. ERP tasks are almost always supported by ERP software systems. ERP is therefore often used to mean ERP software.

Ethernet Ethernet is the leading hardware communications standard for local networks. Originally invented and patented by Robert M. Metcalfe in the 1970s, a significantly modified version is now part of the IEEE 802 specification.

European Network Exchange (ENX) ENX is a European-wide network platform for the automobile industry. Certified by the VDA (Automobile Industry Association) (http://www.enxo.com).

Extranet An extranet is a closed data network, based on Internet technology, that can be used from different locations for communicating with business partners and customers. Every participant/user of an extranet requires specific access permission.

File sharing In this book, the exchange of data between client computers without using a server system. This is also used as a synonym to describe systems used for swapping music, films, and other documents via the Internet.

Firewall Described as a network component through which a secure network is coupled to an insecure network. The task of a firewall is to raise the security in a company network through a number of technical mechanisms. A firewall presents the only access point of a company network to an open network (usually the Internet). It usually consists of several hardware and software components that are individually configured according to user requirements. Concentrating the access to a single network component simplifies the security management as well as the monitoring and control functions. Firewalls are either operated by the company themselves or can also be provided by a third-party company as part of a managed services package.

Flat rate A tariff model for data or speech access to a network system that is independent of usage time or traffic volume. A fixed base price covers all possible usage.

Forwarding equivalence class (FEC) Describes groups that are handled in the same manner.

Forwarding information base (FIB) A router-internal database used to control the forwarding of data packets.

Frame relay A high-speed, connection-oriented, packet-based data communications protocol that transfers information using variable-length data packets. These blocks of data are called frames. Frame relay and IP VPNs have a complex relationship with each other. On the one side, frame relay and IP VPNs compete with each other, but on the other side they can complement each other. Many providers offer transition services.

Frames per second (fps) Pictures per second; term is used as a measurement quantity for video transmission.

Fudge factor A term introduced by Gartner that allows cost evaluation of noncompany use of the company IT/TC resources (e.g., private Internet usage or video games).

G.113 An ITU guideline for speech transmission.

G.114 An ITU recommendation for the transfer time of speech traffic.

G.711 pulse-code modulation G.711 is an ITU standard for digitizing speech information. In this type of modulation, discrete segments of an analog signal are quantized into time and value discrete binary signals. In speech transmission, pulse-code modulation is used to convert an analog speech signal into a digital signal. To do this, the analog signal is sampled 8,000 times a second and converted into an 8-bit value, so that a sample value is obtained every 125 μs. The resulting transmission speed is 64 Kbps and a speech frequency of 4 kHz can be transferred.

G.722, G.723, G.729 ITU recommendations for different coding systems.

Gatekeeper A central control element in H.323 networks with the following tasks: routing of signal information; translation of addresses, telephone numbers, and IP addresses; traffic metering for charging purposes; and bandwidth management to assist QoS.

Gateway The hardware and software necessary to connect different networks to each other. The task of a gateway is to transfer messages from one network to another and to perform any necessary translation of the communication protocols.

General Packet Radio Service (GPRS) An extension of the GSM mobile radio standard for fast data transfer. GPRS is packet oriented, which means that the data is converted to packets by the sender and the receiver reassembles the packets when they are received. GPRS technology makes it possible to bundle all eight GSM channels to achieve a data rate of 171.2 Kbps. In practical operation in German, Austrian, and Swiss data networks, this is limited to a maximum of 53.6 Kbps. Activation of GPRS creates a permanent connection to the target system ("always-on" operation). The radio bandwidth is only used when data actually needs to be sent.

Global System for Mobile Communications (GSM) A worldwide standard for mobile digital radio transmission. GSM mobile radio networks are also described as second generation (2G) networks.

GSM FR A GSM full-rate speech codec. Used in GSM networks among others.

H.323 H.323 is an international ITI standard for speech, data, and video communication over a packet-oriented network. H.323 was developed for the transmission of multimedia applications and forms the basis of VoIP. The H.323 standard consists of several protocols for signaling, for exchanging information about device capabilities, for connection control, for exchange of status information, and for data flow control.

Headset A combined headphone/microphone device. This is the most often used system for VoIP on PCs.

Help desk The help desk (also known as a service desk) is the main contact point for users having questions and problems relating to their IT/TC workplace. The tasks of the help desk include: user and general support; acquisition and classification of fault information; immediate assistance; remote maintenance; and monitoring of servicing operations and fault repair.

High-speed circuit-switched data (HSCSD) An extension of the GSM mobile radio standard that provides faster data transfer. Bundling of several channels allows a theoretical data transfer rate of up to approximately 171 Kbps. The practical usable bandwidth for a device (cell phone) that is not in motion is approximately 30 Kbps. If the device is in motion then a rate of 10 Kbps is realistic.

Home office A workplace in a domestic environment, used either consistently (teleworker) or occasionally (for widely spread groups of employees such as those in field services or management).

Hop A network measurement unit. It reflects the passage through a single router. A distance of two hops means that the packet has passed through two routers in its journey.

Hosted services A form of outsourcing, closely related to the ASP model. From their own computing center, the service provider makes the customer applications available. The hosted services often include Web hosting (among others).

Hub and spoke In WANs: A networking architecture model with star-shaped connections from the central system to the branch offices.

Hype Short-term, exaggerated interest in a technology, product, or person. This is often provoked or promoted by marketing activities.

Hypertext Markup Language (HTML) Language used for defining the content of Web pages in the World Wide Web.

Hypertext Transfer Protocol (HTTP) Protocol used for transferring data using the TCP/IP protocol stack. The main area of application is the transfer of Web site information between a Web server and a Web browser on the client system.

Information Technology Infrastructure Library (ITIL) The ITIL is a manufacturer-independent collection of best practices for IT service management. Originally founded as an initiative of the British government at the end of the 1980s, the concept has been developed further. ITIL is seen as the de facto standard for IT service management (http://www.itil.co.uk).

Infotainment An invented word formed from *information* and *entertainment.* Among other uses, the word describes a possible source of income for UMTSnetworks.

Instant messaging A form of communication, originally developed for private applications, that allows users to send messages (and attached data) directly to other users, thus making an online real-time discussion possible. Current implementations also have mechanisms for determining if the desired discussion partner is currently online or not. There are open instant messaging systems, mostly provided by Internet service providers and which should not be used for business purposes because of security risks (lack of encryption, unmonitored data transfer, and so forth), and also closed systems operated by businesses especially for their internal company communications.

Institute of Electrical and Electronics Engineers (IEEE) The IEEE is an international organization of electrotechnology and computing engineers, similar to the German VDE (Verband der Elektrotechnik Elektronik Informationstechnik e.V.). The IEEE organizes professional conferences and forms committees for standardizing hardware and software technology. Among other standards, they are known for implementing the IEEE 802 standard. The IEEE norms committee also develops network standards. The best known of these standards is the 802.11 standard for wireless LAN.

Integrated Services Digital Network (ISDN) ISDN was originally conceived as a communications network for voice traffic. The digital transmission allows text, graphics, and voice data to be handled in the same manner. ISDN is link based in a similar manner to the analog long-distance network. ISDN is available as a basic connection and also as a primary multiplex connection. The basic connection provides the user with two basic channels (B-channels), each of which is at 64 Kbps, and also a control channel, the so-called D-channel, with 16 Kbps. Basic connections are available as multiple device connections (point-to-multipoint) for connection of up to eight ISDN end devices or system connections (point-to-point) for connecting a single telecommunications system, for example, a telephone exchange. The primary multiplex connection has 30 usable channels. It is only available as a primary equipment connection and is mainly used by large companies for connecting telephone systems.

International Telecommunication Union (ITU) A worldwide organization in which governments and private telecommunications companies coordinate the construction and operation of telecommunications networks. The ITU is responsible for the regulation, standardization, coordination, and development of international and national telecommunications. Originally founded in Paris in 1865 by 20 countries, it was made into a branch of the United Nations in 1947, and is now based in Genf. There are three sectors (offices) in the ITU, each of which has its own director: radio communications (BR), telecommunications standardization (TSB), and telecommunications development (BDT). ITU-T-Standards (or ITU/TS Telecommunications Standards) are adopted on the basis of CCITT recommendations.

Internet Engineering Task Force (IETF) Along with the Internet Research Task Force (IRTF), the IETF is one of two workgroups with the Internet Architecture Board (IAB). It is an international association of network technicians, manufacturers, and users that is responsible for Internet standardization proposals. The IETF currently operates in nine areas: applications (APP); Internet services; IP: Next generation (IPNG); network management (MNT); operations (OPS); routing (RTG); security; transport services; and user services (USV). The association was founded in 1986. In contrast to the IRTF, the IETF focuses more on the short-term development of the Internet. There are currently more than 80 workgroups with more than 700 members.

Internet Protocol (IP) A transmission standard used in the public Internet and in most company networks. IP transfers text, Web content, and also audiovisual information in the form of numerous data packets (packet forwarding). The arrival of all packets at the recipient, in the correct order, is guaranteed by this protocol. A response mechanism to indicate receipt of the packets is not part of IP. This disadvantage is solved by the Transmission Control Protocol (see TCP), which runs on top of IP.

Internet service provider (ISP) A company providing access to the Internet.

Internet telephony A low-cost, but potentially problematic, variation of telephony that has been known for a number of years. The system uses the public Internet for transmission of speech information. The consequences: possibly varying quality, even as far as pauses or dropouts. PCs or notebooks with headsets, and increasingly with analog telephone adaptors, are typically used as the end devices. Both partners must already be on line in order to establish communication. The costs for conversations within the Internet structure are typically limited to (usually low) data volume costs.

Intranet An intranet is a company-internal, access-protected network, usually based on Internet technology. The core of intranet applications is usually WWW services.

Intrusion detection system (IDS) A security technology using a continuous analysis of network traffic and recognition of anomalies to provide early detection of security attacks.

IP VPN A VPN based on the Internet protocol (this is normally the case at present).

IPsec; IPSec A security architecture for secure communication over IP networks. The protocol is supposed to guarantee trustworthiness, authenticity, and integrity of transmissions. In addition, it should provide protection against so-called "replay attacks," meaning that an attacker cannot trick the remote device into repeating a given action by simply repeating a previously recorded set of data. The core document relating to IPsec is RFC 2401. In contrast to other encryption protocols, such as SSH, the IPsec system functions at the network layer (layer 3) of the OSI reference model.

Java An object-oriented, platform-independent programming language developed by SUN Microsystems. Java is not related to JavaScript.

JavaScript A scripting language for use in the Internet developed by Netscape Corporation. JavaScript programs are embedded in HTML pages and executed on the client computer. JavaScript is not related to Java.

Jitter In transmission technology, jitter is described as an abrupt, unwanted change in a signal characteristic (amplitude and/or frequency). Jitter has a negative effect on the perceived quality of VoIP.

Label Distribution Protocol (LDP) A signaling protocol, defined by the IETF, used for establishing label-switched path (LSP) connections in MPLS networks.

Label edge router (LER); E-LSR A router at the edge of an MPLS network that classifies IP packets and adds a corresponding label to each packet. (See also LSR.)

Label-switched path (LSP) A connection path within an MPLS network.

Label-switched router (LSR) MPLS routers offer additional functionality and are described as so-called LSRs (label-switched routers). These are differentiated into edge LSRs (E-LSRs) and core LSRs (C-LSRs). A C-LSR performs the packet forwarding in the core of the MPLS network.

LAN–LAN coupling A connection between two local-area networks, typically through a VPN.

Latency A synonym to describe a delay time. In general, this is the time interval between the end of an event and the beginning of a response to this event.

Leased line Permanent line. A permanent communications link provided for a fee.

Letter of intent (LOI) An LOI is a written declaration of some type of intention. Often used when companies begin some form of long-term business collaboration. Depending on the form of the particular LOI, some form of commitment to finalizing a contract exists or, if a contract is not finalized, some form of financial compensation. If the LOI is weakly structured, then it may in fact not be binding.

Local-area network (LAN) A class of fast computer network, with mainly decentralized communications control, that allows a number of equally authorized users (stations) to cooperatively use the bandwidth of a fast transmission medium over a (usually) limited geographical range. The range of a LAN is often limited to one or more buildings at the same site, and separations of more than 10 km are rare. The technology available up to now allows transmission rates of up to approximately 1,000 Mbps.

Local loop Describes the connection between the forwarding center, or POP server, and the customer transfer point. In the case of the public telephone network, this is the area between the local forwarding center and the forwarding point in the home; this can be the access point or also the telephone socket.

Location-based services Services that provide location-relevant information, such as navigation. Usually used in the context of mobile radio networks.

Loss The loss of a data packet during transmission.

Malware A general term for all software having the potential to cause damage, for example, viruses, trojans, worms.

Man-in-the-middle attack A form of network attack in which the attacker places himself between sender and receiver and manipulates the data traffic for his own purposes.

Managed network services/managed data network services Describes the services offered by a telecommunications company. Managed network services cover the planning, construction, administration, and maintenance of large networks. The scope of the services typically extends as far as the routers, forming the customer interface to the larger system. Complete IP-based corporate networks can be operated on behalf of the customer over these networks.

Managed security services Outsourcing of the operation of security systems. Main focus: assisted operation of firewalls.

Managed services A form of outsourcing with the main focus on assisted operation of an infrastructure. (*See also* Managed network services.)

Market intelligence Strategic observation of a market. Also used as a synonym for *competitive intelligence.*

Media access control (MAC) address A unique hardware address belonging to every device in a network.

Messaging The exchange of messages between one or more systems.

Micro-multinationals A description of start-up companies that use offshoring from the very beginning of their existence.

Monitoring In general, the continuous observation of an action or process through the use of some sort of technical assistance, such as special monitoring software.

Moving Picture Experts Group (MPEG) A group of experts dedicated to the standardization of video compression and related areas. The group often extends and improves ideas from the ITU. Since the first meeting in 1988, the group has grown to approximately 360 members from various companies and academic institutions and has helped to establish many standards.

Multicast A form of transmission in a computer network, from one point to a group of destinations. This is also known as a *multipoint connection.* Multicasting allows messages to be sent simultaneously to several participants or a closed group of participants.

Multimode fiber A fiber-optic conductor with a large diameter that also allows the transmission of nonaxial radiation.

Multiprotocol Label Switching (MPLS) MPLS allows every application to be allocated a different level of service. Telephone conversations, images, sound from videoconferences, and important applications (such as ERP systems) have higher priority. Other applications, such as e-mail or downloads, must wait when the network bandwidthis limited. To guarantee this, every MPLS data packet receives a labeldefining this priority. The system then searches for the best path through the router network for every data packet. This avoids the time-consuming forwarding of packets in a hop-by-hop routing system. This allows for a rapid, continuous connection to be established. MPLS is standardized by the IEFT.

Nearshore outsourcing; nearshoring *See* Outsourcing.

Network address translation (NAT) A process occurring in computer networks in which private IP addresses are translated into public IP addresses. NAT is necessary because there is a shortage of open IP addresses and this forces the use of local IP addresses. NAT also helps with data security because the internal structure of the network remains hidden from the outside world.

Network operating center (NOC) A facility performing central operating tasks within a network (e.g., monitoring, control, fault recognition, user support). This usually occurs within a provider network.

Next generation network (NGN) An inconsistently used term that describes an integrated network of the "next generation."

Offshoring; offshore outsourcing *See* Outsourcing.

On-demand computing A term coined by IBM that refers to the provision of computing resources, according to requirements, when required. This concept is also described as *organic IT*.

Open Shortest Path First (OSPF) A dynamic routing protocol.

Outsourcing A general term for using an external provider to perform company activities and/or services. Here are the most important outsourcing terms, based on the article "15 Years of Outsourcing Research: Systemization and Lessons Learned" from *Information Management Concepts and Practical Strategies* (2004) by Holger von Jouanne-Diedrich, who has isolated more than 30 different outsourcing terms.

- *Onshore outsourcing* occurs when the outsourcing service provider is in the same country as the outsourcing customer.
- *Nearshore outsourcing* occurs when the outsourcing activities include one or more outsourcing providers from neighboring countries.
- *Offshore outsourcing* is the term used when the geographical separation between the outsourcing customer and the outsourcing provider is the greatest (also known as *offshoring* or *farshoring*). India is a typical example of this.
- *Bestshoring* comes from the U.S. company EDS, which has a long tradition as an outsourcing service provider. Under this term, EDS includes the attempt to provide the outsourcing customer with "the best tool for the job," that is, doing the job in the country where it can best be done.

- *Rightsourcing* is used by some outsourcing service providers who are of the opinion that they are especially good at tailoring their services to suit the individual requirements of the customer.
- *Internal outsourcing* refers to the situation in which outsourcing contracts are awarded to daughter companies.
- *External outsourcing* is the opposite of internal outsourcing.
- *Tactical outsourcing* occurs when a company switches between internal and external outsourcing. This is the term used by Silicon.de when the outsourcing customer allows relatively small tasks to be performed externally and requires internal and external outsourcing service providers to compete with each other to provide the best offer.
- *Strategic outsourcing* occurs when an outsourcing provider assumes entire business processes of the outsourcing customer.
- *Transforming outsourcing* results in significant changes in the business of the outsourcing customer. Outsourcing using this variant causes massive upheaval in the business processes of the customer.

Other terms focus on the particular range of tasks to be performed by the service provider. These include *selective outsourcing*, outtasking, smart outsourcing, and best-of-breed outsourcing. In this case, only a clearly defined task or portion of a task is outsourced. Yet other terms focus on the explicit description of the task to be performed:

- *Desktop outsourcing*, in which the outsourcing provider assumes tasks such as installation or repair of the desktop computers;
- Midrange outsourcing;
- Mainframe outsourcing;
- Network outsourcing;
- Infrastructure outsourcing;
- Application outsourcing;
- Information technology outsourcing.

Business application outsourcing is a term used by Accenture to describe the development and management of software by an outsourcing provider. Accenture has also coined the term *cosourcing*, in which employees of the outsourcing provider assume key positions in the company of the outsourcing customer. *On-demand sourcing* occurs when the outsourcing customer only uses the services that they actually require. *Precision sourcing* is another term invented by a consulting company.

While many terms have been created under company-specific circumstances and have limited application for general use, others have become firmly established, for example, the term used to describe the allocation of entire business processes to an external company: *business process outsourcing* (BPO).

Terms have also been invented to describe the process of returning outsourced tasks back to the customer, for example, if the outsourcing attempt was not successful, the return is referred to as *insourcing* or *backsourcing*.

Outtasking In contrast to outsourcing, only individual tasks (or partial tasks) are given to an external provider, not entire business processes. Planning and control remain with the customer, who is then somewhat less dependent on the service provider.

Packet loss rate The percent of packets lost over a particular transmission path.

Patch Corrections or improvements to existing software.

Patch panel A plug panel used for connecting elements of speech and data networks.

Peak signal-to-noise ratio (PSNR) PSNR is the highest value of the relationship of signal to noise in images and is a parameter used for measuring the quality of video data or the codec used for compression.

Peaks Sudden high values in one or more characteristics of a signal (e.g., when a connection is overloaded).

Personal digital assistant (PDA) A small, handheld device providing common daily functions such as a calendar, diary, or notebook. Describes the smallest form of a portable computer.

Performance A measurement of the speed, processing capacity or functionality of a system

Point of presence (POP) Defines the place at which access to the network exists, for example, the access point to the Internet with an Internet provider or the place where the carrier for the wider network connects to the local provider. The connection between the customer location and the POP is called the *local loop*.

Portable Document Format (PDF) A widely used data format from Adobe Corporation for exchanging or archiving documents.

Portal *See* Company portal.

Postpaid In contrast to prepaid telecommunications, telecommunications services that are paid for after they have been provided. Identification of the customer is a compulsory prerequisite of this process.

Precision sourcing A term coined by consultants to describe selective outsourcing.

Prepaid A payment method for telecommunications services in which a specific credit is established, in the form of a credit note or similar, before services are provided and this is then used to pay for the use of the system. This process does not require the customer to be identified for invoicing purposes. Prepaid is the most common form of accounting for public WLAN systems.

Primary multiplex connection *See* ISDN.

Private automatic branch exchange (PABX) *See* PBX.

Private branch exchange (PBX) An English term for a telephone exchange system to which one or more end devices, so-called extensions, are connected and which is connected to the public telephone network by one or more main connections.

Provider A provider is usually a company offering some form of network access or services.

Public switched telephone network (PSTN) The public telephone network.

Pulse Code Modulation (PCM) *See* G.711.

Q.931 This is a protocol, standardized by the ITU, for sending signals over the D-channel of European ISDN connections that administrate creation and control of connections.

Quality of service (QoS) QoS is an agreed-on parameter for defining the quality of a service in some way.

Real-Time Control Protocol (RTCP) A control protocol, designed as an extension to RTP, to allow the guarantee of certain characteristics of the data transfer (e.g., guaranteed bandwidth or priority).

Real-Time Transport Protocol (RTP) This protocol is part of H.323. It lies in the application layer and can support network-based video or audio communication. The RTP protocol is based on an end-to-end connection and supports both IP multicast and unicast connections. It detects and corrects missing, duplicated, or wrongly sequenced data packets using a 16-bit sequence number.

The protocol uses a timestamp, defined by the RTP protocol being used, to help with synchronizing audio or video traffic.

Remote access service (RAS); remote dial-up A data communications service that allows remote access from a distant workplace to the company network and services. This was previously done by direct dialing a special server (RAS server) via a modem or ISDN, but these days, connections between remote clients and company networks are done using a VPN over the public Internet.

Remote survivability This is the functional capability of the systems (e.g., VoIP telephone systems) in a branch office when the data connection to the central system drops out.

Request for comment (RFC) A collection of technical and organizational Internet-related documents defining various standards. When first published, they are made available for public discussion, in the true sense of the name, but RFCs are still called RFCs when general acceptance and use have turned them into a standard.

Request for proposal (RFP) A document defining the conditions for making an offer for products, services, or for suggesting a standard.

Resource Reservation Protocol (RSVP) A protocol for reserving resources in routers. The RSVP protocol makes it possible to create end system connections on top of the normally connectionless IP protocol and thus define different classes of traffic (class of service).

Return on investment (ROI) This is an Anglo-American term for the profitability of an investment. This value provides information about the capital interest received on an investment over a given period. ROI represents the profit on the invested capital for each unit in the system. ROI is one of the most frequently used characteristics in the financial analysis of systems. It is often used as a measure of the performance and profitability of entire companies, business areas, or individual investments.

Rich media communications Multimedia communications achieved through the coupling of speech, data, and video traffic.

Roaming In general, describes a mobile device moving from one base station to another. This usually occurs without the connection being broken or the user noticing any changes in the connection. Roaming can also mean changing from one network to another.

Router A forwarding computer for connecting different networks or network segments. The router forwards data packets according to the address information contained in the packet and the routing tables stored within the device (*routing*). Routers are usually specialized devices. It is possible, however, to use industry standard computers as routers. In this case the routing is implemented in software. In the Soho segment, simple routers—often combined with WLAN and switch functionality—are used for connecting to the Internet. This type of router can also be used to support VPN functionality by connecting teleworker workplaces.

Routing table For creating an end-to-end connection between end devices in a packet-forwarding network, for instance, to transmit the data packet from sender to recipient, both end stations (the sending station and the receiving station) must be uniquely identifiable by their network address. The routers lying between the end devices forward the data traffic through the network to the target device (i.e., from the first router to the second, to the third, and so forth) according to the routing tables stored in each router. The content of the routing tables in each router defines the forwarding information for incoming packets. The forwarding occurs on the basis of the address information in the data packet and the information found in the routing table that defines the path to take in order to reach the target address.

S2M *See* ISDN.

Scoring The process of accumulating points. It is used for various analysis methods and is used in this book as part of the process for choosing a supplier.

Secure Real-Time Transport Protocol (SRTP) An Internet draft proposal; SRTP offers security, confidentiality, and authentication/integrity for RTP/RTCP packets. (*See also* RTP.)

Secure Socket Layer (SSL) SSL describes a transmission protocol, originally developed by Netscape Corporation, with which encrypted communication is possible through the use of HTTP tunneling. The most important advantage of SSL is the ability to implement higher level protocols based on SSL. This guarantees the independence of applications and systems. SSL is used for the transmission of encrypted Web sites via HTTPS.

Security Network security.

Service-level agreement (SLA) This is a two-way contractual agreement between a network provider and a customer, in which the quality of the performance of the system is explicitly specified. The most important aspects of such quality agreements are bandwidth, availability, network capacity, and network quality. In addition to the technical parameters, the quality of service, the technology, and the technology used to measure the service and monitor it, the dropout times, the reaction times, and repair times all play a role. An SLA covers all services, networks, components, and connections: speech, data and Internet services, basic and value-added services, fixed and mobile radio networks, MAN and WAN, and fixed, directional radio, and end-to-end connections. Although currently there are no standards defined for SLAs, every service provider can include whatever offers and services they wish in the SLA. An SLA contract should contain definitions of the object(s) it relates to, the duration of the contract, the terms of cancellation, the technical performance to be provided, the manner in which faults are handled, the penalties incurred if specifics terms are not fulfilled, and the change management procedures.

Session Initiation Protocol (SIP) A signaling protocol that can create, modify, and end sessions between two or more participants. The HTTP-based, text-oriented protocol permits the transmission of real-time data over a packet-based network. In terms of its functionality, the SIP protocol is similar to the H.323 protocol and can transmit interactive communications services, including speech, over an IP network. SIP information can be transmitted using TCP or UDP. SIP has an open structure and allows features such as transmission of caller identity or call forwarding over IP networks.

Shared hosting A hosting business model in which several customers use a server environment together.

Short message service (SMS) A service for the transmission of short text messages over the signaling channel of a GSM network. Can be used for messaging and telemetry applications. SMS services are now also available in the normal public telephone network.

Simple Mail Transfer Protocol (SMTP) SMTP belongs to the family of standard TCP/IP protocols and controls the sending of e-mails in computer networks. The SMTP standard is defined in RFC 2821 (http://www.faqs.org/rfcs/rfc2821.html). Despite the widespread and intensive use of SMTP, it has serious weaknesses. There are no security mechanisms, especially for authentication (authenticity and nonrepudiation; the determination and classification of the identity of the sender), integrity (ensuring that no changes to the content occur during transmission), and encryption. Even extra mechanisms that can help with these problems are seldom and inconsistently used.

Single sourcing A procurement strategy in which a company purchases goods and services from one supplier only. In multiple sourcing, several suppliers are considered for the purchase of a product or service. The advantages of single sourcing generally lie in the saving of transaction and handling costs and also in the use of quantity discounts. The main advantage of multiple sourcing is low dependency on a single supplier.

Spam Normal term for unwanted advertising e-mails. These are being sent in ever increasing numbers, due to the weaknesses of the SMTP protocol used for e-mail. Other instances of spam include clogging up discussion forums (previously Usenet groups) with unwanted advertising messages or search machine spam (where the search machine hit list is manipulated through specially prepared Web sites).

Streaming audio/video The continuous transmission of data, typically audio and video data, over a network, for example, a company intranet or the public Internet. The transmission should occur without delays, if possible. For this reason, the data is displayed/played by the end application during the transfer process, in contrast to downloads where the audio and video data are only readable once the transmission is finished. Streaming requires a special streaming server that sends data to a streaming client (e.g., Windows Media Player or RealPlayer) and which reacts to control commands from the client machine. Some possible application areas for audio/video streaming include: Internet radio; audio and videoconferencing; video on demand, for example, in the form of an online video shop or by calling up prestored videos of conferences or concerts; and business TV.

Subscriber The customer subscribing to and using a service, in this case network services.

Switch A network component, with switching functionality, for connecting several computers or network segments.

Synchronization The process of two things occurring at the same time.

TC Telecommunications.

TCP/IP *See* TCP.

Telemetry Telemetry, or remote measuring/remote control engineering, encompasses measurement and control systems that function over a telecommunications link, for example, status transmission and querying of server system functions.

Teleworking Term that covers a range of different types of working. The common element is the use of IT and data communication to assist with performing the work. The following types of working are usually distinguished: home teleworking, alternating teleworking, and mobile teleworking.

The definition can also be somewhat extended to include on-site teleworking and working in a telecenter. In home teleworking, the employee performs all the work in his own home. He has no workplace on the company premises. Alternating telework is the most common variation of telework. Here, the employee works both at home and on the company premises, normally one day at a time. This is often combined with desk sharing. Here, two employees share a desk, or there is a pool of desks that are not allocated to any particular person. The company also saves space and money with this system.

Advisors, field employees, service personnel, and those in related occupations mainly do mobile telework. In this variant, the work is performed at a number of changing locations (e.g., at the customer site or at home). Secure and reliable remote access to the company IT infrastructure is usually the dominant infrastructure problem in this case. On-site telework is performed on the premises of another company. Business advisors, for example, practice this type of telework.

An increasingly rare form of telework that is also becoming less and less important is telework in so-called telecenters. Here, the employees do not work on the company premises but in an office near to the company site that is serviced by the company infrastructure (phone and network access). This type of telecenter can be maintained and operated by a single company, but it is often a type of corporate gathering place for employees of many different companies. In this case, the companies share the operating costs of the telecenter. In the 1980s, this type of telecenter was promoted in order to assist regional development. It is almost an irony of history that similar concepts are now resurfacing in the concept of offshore outsourcing.

The advantages and disadvantages of teleworking are still somewhat controversial in the broader working environment.

Thin client Thin clients are network-based workplace computers whose functionality is reduced to input and output operations. The operating system and the applications are stored in central servers.

TIME An acronym and general term for the following markets: Telecommunications, Information technology, Media, Entertainment. During the New Economy boom, the combining of these markets into the future TIME market was seen as an important future development. Current developments indicate that this is actually occurring, albeit somewhat slower than anticipated.

Time-out A definition of how long a system will wait for an event to occur before a procedure is aborted.

Toll-free number A telephone number that is free for the customer (e.g., 800). This type of number is often used for customer support services.

Total cost of ownership (TCO) A term, originally propagated by the Gartner Group, used for determining the total costs associated with a potential IT or TC investment. In addition to the direct costs of purchasing a new object or service, indirect costs are also calculated, for example, such as costs associated with possible dropouts. The TCO concept is discussed in detail in Chapter 5.

Traffic Describes the data traveling through a network.

Transmission Control Protocol (TCP) TCP is a transport layer protocol for use in packet forwarding networks. This connection-oriented protocol is implemented on top of the IP protocol, supports the transport layer functions, and creates a guaranteed connection between the endpoints before any data is transmitted. The data in the higher layers is not modified by TCP. The data is sent in individual packets. The packets receive sequence numbers to ensure that they are reassembled in the correct order at the receiving end.

Triple play Originally: A baseball game tactic. In this book, the strategic orientation of a provider to provide a combined offer of Internet, TV, and telephony.

UMTS Forum The UMTS Forum (http://www.umts-forum.org) was founded in 1996 and is an independent international alliance of approximately 200 organizations, companies, and governmental departments who are developing a comprehensive concept for the introduction and development of third generation mobile telephone networks. The forum concentrates on the preparation of market-oriented recommendations to the regulators, network operators, and service providers as well as stimulating discussion on spreading a new wideband, personalized access to information worldwide.

Unicast Data that is only addressed to a single recipient, in contrast to broadcast and multicast.

Unified messaging; unified messaging service A service that gathers and administrates various types of messages at a single point (e.g., speech, fax, e-mail, and SMS messages). The recipient can retrieve and process her messages from any location and with any suitable end device. A unified messaging system can also automatically distribute messages to particular recipients.

Unified synchronized communications (USC) The integrated use of different communications technologies (speech, data video).

Uniform resource locator (URL) A form of address for Internet data, used mainly in the World Wide Web. The URL format allows unique identification of all documents on the Internet. The URL contains the type of transmission to be used (Gopher, HTTP, ftp, news, and so forth), the address of the target computer containing the data, and the file path of the data on the target computer.

Universal Mobile Telecommunications (UMTS) UMTS is a third generation mobile radio network. UMTS allows packet-based transmission of data up to 2 Mbps. Current implementations are limited to 384 Kbps.

Upstream Describes an outgoing flow of data. In IT, one can describe a flow of data as a sequence of data where the end of the sequence cannot be predicted in advance. (*See also* Downstream.)

User Datagram Protocol (UDP) A transport layer protocol for use in packet forwarding networks. The protocol is based on the Internet Protocol, is connectionless, and, in contrast to TCP, does not guarantee delivery of the packets.

Utility computing With utility computing, the user pays for the performance used. This is either provided in-house, for example, in the company computing center, or (more often) by a service provider. The term, which has been propagated by HP and other suppliers, comes very close to the IBM concept of on-demand computing.

Virtual private network (VPN) A VPN is a lower cost alternative to leased lines. The customer of a network provider does not rent a cable connection but rather books some type of transport service. The customer data is transmitted over the provider network at an agreed level of service quality.

Voice spam Automatic verbal advertising via telephone. Almost always unwanted. With the further expansion of VoIP, there are fears of a massive expansion of this type of annoyance.

Voicemail Voicemail is a speech storage service. This is a replacement for the classical telephone answering machine. The recipient can retrieve the message at any desired time. Apart from the basic functionality of the normal telephone answering machine, voicemail systems also frequently offer other services such as message forwarding and notification functions.

Voice over IP (VoIP) VoIP describes the transmission of speech traffic over an IP network. Although Internet telephony also falls under this general term, VoIP specifically refers to transmission at a defined level of quality that is oriented to the QoS available with normal public telephone systems.

Web content management system *See* Content management system.

Web hosting This is the provision of Web sites and the operation of Web applications on a server owned by the provider as part of a paid service. As the service provider, the Web hosting company makes storage on a server available (a so-called "virtual server") or makes an entire server available (a so-called "dedicated server") for exclusive use. The user can make information available and—depending on the system—can also run her personal applications (e.g., online ordering, customer support). Web hosting services are offered by specialist hosting companies as the main product, but are also available from telcos and

IT service providers as side products in addition to the usual palette of products and services.

White paper Term originally used to describe a status paper defining the position taken by an organization in regard to, say, political questions. During the past few years, the IT and TC industries have also adopted this idea of a paper containing a brief explanation of a topic. The typically 2- to 10-page documents contain an overview of a particular product or technology. The terminology used is usually not particularly technical since the aim is to address a wider audience. In addition to technical white papers in the IT area, there are now also "nontechnical white papers" that deal in detail with the effects on the company of a particular technology or concept.

Wide-area network (WAN) Network that transmits speech or data over long distances. In companies, WANs are typically used for connecting regionally distributed locations. WANs are conceived with regard to the services that are to be offered. A WAN can be link based or packet based.

Wireless LAN (WLAN) Refers to variants of the IEEE 802.11 standard, with different transmission speeds. The idea of WiFi (wireless fidelity) often occurs in conjunction with WLAN. This is a certification process performed by the manufacturers' organization Wireless Ethernet Compatibility Alliance (WECA). This certification confirms the interoperability of WLAN products with regard to the WLAN standards.

About the Author

Thomas R. Koehler is an Internet entrepreneur and leading expert in enterprise networking. He received a degree in business administration from Wuerzburg University (Germany). He teaches computer sciences and competitive intelligence at Ansbach University of Applied Sciences, Germany. His current research focuses on VoIP and convergence strategies for large corporations and incumbent telcos. He is a frequent speaker at international congresses and corporate events. He has written several books, numerous trade magazine articles, and a television broadcast on internetworking. His e-mail address is mail@thomaskoehler.de.

Index

Recent Titles in the Artech House Telecommunications Library

Vinton G. Cerf, Senior Series Editor

Access Networks: Technology and V5 Interfacing, Alex Gillespie

Achieving Global Information Networking, Eve L. Varma et al.

Advanced High-Frequency Radio Communications, Eric E. Johnson et al.

ATM Interworking in Broadband Wireless Applications, M. Sreetharan and S. Subramaniam

ATM Switches, Edwin R. Coover

ATM Switching Systems, Thomas M. Chen and Stephen S. Liu

Broadband Access Technology, Interfaces, and Management, Alex Gillespie

Broadband Local Loops for High-Speed Internet Access, Maurice Gagnaire

Broadband Networking: ATM, SDH, and SONET, Mike Sexton and Andy Reid

Broadband Telecommunications Technology, Second Edition, Byeong Lee, Minho Kang, and Jonghee Lee

The Business Case for Web-Based Training, Tammy Whalen and David Wright

Centrex or PBX: The Impact of IP, John R. Abrahams and Mauro Lollo

Chinese Telecommunications Policy, Xu Yan and Douglas Pitt

Communication and Computing for Distributed Multimedia Systems, Guojun Lu

Communications Technology Guide for Business, Richard Downey, Seán Boland, and Phillip Walsh

Community Networks: Lessons from Blacksburg, Virginia, Second Edition, Andrew M. Cohill and Andrea Kavanaugh, editors

Component-Based Network System Engineering, Mark Norris, Rob Davis, and Alan Pengelly

Computer Telephony Integration, Second Edition, Rob Walters

Customer-Centered Telecommunications Services Marketing, Karen G. Strouse

Deploying and Managing IP over WDM Networks, Joan Serrat and Alex Galis, editors

Desktop Encyclopedia of the Internet, Nathan J. Muller

Digital Clocks for Synchronization and Communications, Masami Kihara, Sadayasu Ono, and Pekka Eskelinen

Digital Modulation Techniques, Fuqin Xiong

E-Commerce Systems Architecture and Applications, Wasim E. Rajput

Engineering Internet QoS, Sanjay Jha and Mahbub Hassan

Error-Control Block Codes for Communications Engineers, L. H. Charles Lee

Essentials of Modern Telecommunications Systems, Nihal Kularatna and Dileeka Dias

FAX: Facsimile Technology and Systems, Third Edition, Kenneth R. McConnell, Dennis Bodson, and Stephen Urban

Fundamentals of Network Security, John E. Canavan

Gigabit Ethernet Technology and Applications, Mark Norris

The Great Telecom Meltdown, Fred R. Goldstein

Guide to ATM Systems and Technology, Mohammad A. Rahman

A Guide to the TCP/IP Protocol Suite, Floyd Wilder

Home Networking Technologies and Standards, Theodore B. Zahariadis

Implementing Value-Added Telecom Services, Johan Zuidweg

Information Superhighways Revisited: The Economics of Multimedia, Bruce Egan

Installation and Maintenance of SDH/SONET, ATM, xDSL, and Synchronization Networks, José M. Caballero et al.

Integrated Broadband Networks: TCP/IP, ATM, SDH/SONET, and WDM/Optics, Byeong Gi Lee and Woojune Kim

Internet E-mail: Protocols, Standards, and Implementation, Lawrence Hughes

Introduction to Telecommunications Network Engineering, Second Edition, Tarmo Anttalainen

Introduction to Telephones and Telephone Systems, Third Edition, A. Michael Noll

An Introduction to U.S. Telecommunications Law, Second Edition, Charles H. Kennedy

IP Convergence: The Next Revolution in Telecommunications, Nathan J. Muller

LANs to WANs: The Complete Management Guide, Nathan J. Muller

The Law and Regulation of Telecommunications Carriers, Henk Brands and Evan T. Leo

Managing Internet-Driven Change in International Telecommunications, Rob Frieden

Marketing Telecommunications Services: New Approaches for a Changing Environment, Karen G. Strouse

Mission-Critical Network Planning, Matthew Liotine

Multimedia Communications Networks: Technologies and Services, Mallikarjun Tatipamula and Bhumip Khashnabish, editors

Next Generation Intelligent Networks, Johan Zuidweg

Open Source Software Law, Rod Dixon

Performance Evaluation of Communication Networks, Gary N. Higginbottom

Performance of TCP/IP over ATM Networks, Mahbub Hassan and Mohammed Atiquzzaman

Practical Guide for Implementing Secure Intranets and Extranets, Kaustubh M. Phaltankar

Practical Internet Law for Business, Kurt M. Saunders

Practical Multiservice LANs: ATM and RF Broadband, Ernest O. Tunmann

Principles of Modern Communications Technology, A. Michael Noll

A Professional's Guide to Data Communication in a TCP/IP World, E. Bryan Carne

Programmable Networks for IP Service Deployment, Alex Galis et al., editors

Protocol Management in Computer Networking, Philippe Byrnes

Pulse Code Modulation Systems Design, William N. Waggener

Reorganizing Data and Voice Networks: Communications Resourcing for Corporate Networks, Thomas R. Koehler

Security, Rights, and Liabilities in E-Commerce, Jeffrey H. Matsuura

Service Assurance for Voice over WiFi and 3G Networks, Richard Lau, Ram Khare, and William Y. Chang

Service Level Management for Enterprise Networks, Lundy Lewis

SIP: Understanding the Session Initiation Protocol, Second Edition, Alan B. Johnston

Smart Card Security and Applications, Second Edition, Mike Hendry

SNMP-Based ATM Network Management, Heng Pan

Spectrum Wars: The Policy and Technology Debate, Jennifer A. Manner

Strategic Management in Telecommunications, James K. Shaw

Strategies for Success in the New Telecommunications Marketplace, Karen G. Strouse

Successful Business Strategies Using Telecommunications Services, Martin F. Bartholomew

Telecommunications Cost Management, S. C. Strother

Telecommunications Department Management, Robert A. Gable

Telecommunications Deregulation and the Information Economy, Second Edition, James K. Shaw

Telecommunications Technology Handbook, Second Edition, Daniel Minoli

Telemetry Systems Engineering, Frank Carden, Russell Jedlicka, and Robert Henry

Telephone Switching Systems, Richard A. Thompson

Understanding Modern Telecommunications and the Information Superhighway, John G. Nellist and Elliott M. Gilbert

Understanding Networking Technology: Concepts, Terms, and Trends, Second Edition, Mark Norris

Videoconferencing and Videotelephony: Technology and Standards, Second Edition, Richard Schaphorst

Visual Telephony, Edward A. Daly and Kathleen J. Hansell

Wide-Area Data Network Performance Engineering, Robert G. Cole and Ravi Ramaswamy

Winning Telco Customers Using Marketing Databases, Rob Mattison

WLANs and WPANs towards 4G Wireless, Ramjee Prasad and Luis Muñoz

World-Class Telecommunications Service Development, Ellen P. Ward

For further information on these and other Artech House titles,
including previously considered out-of-print books now available through our
In-Print-Forever® (IPF®) program, contact:

Artech House
685 Canton Street
Norwood, MA 02062
Phone: 781-769-9750
Fax: 781-769-6334
e-mail: artech@artechhouse.com

Artech House
46 Gillingham Street
London SW1V 1AH UK
Phone: +44 (0)20 7596-8750
Fax: +44 (0)20 7630-0166
e-mail: artech-uk@artechhouse.com

Find us on the World Wide Web at:
www.artechhouse.com